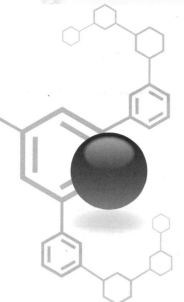

A Primer in Prebiotic Chemistry

Albert C. Fahrenbach

Henderson J. Cleaves II

OXFORD

UNIVERSITY PRESS

Tables

Figures

Preface

"The long saga of humanity's quest to solve the riddle of the origin of life is filled with scientists who thought they were on the brink of solving the great mystery, only to see their discoveries and contributions washed away by the acid test of scientific scrutiny."

From *A Brief History of Creation*, Mesler and Cleaves.

This book aims to give a general introduction to the topic of prebiotic chemistry, pitched at a level accessible to 3rd/4th year undergraduates, as well as first-year graduate/honours students. Prebiotic chemistry is a cutting-edge contemporary topic in chemistry that brings together knowledge from bio-, organic, physical, geo-, and astrochemistry. Many universities in the US, Europe, Asia, Latin America, and Australia offer courses at this level in astrobiology that will benefit from this book.

Though the origins of life have been the topic of scientific study for well over 50 years, and astrobiology is highly topical, most texts focus on astronomy, microbiology, or planetary science—core/central chemical concepts often fall to the background. Prebiotic chemistry offers a logical way to make this topic more relevant to the chemical sciences and will be an excellent part of foundational chemistry courses which will help connect chemistry with other science disciplines.

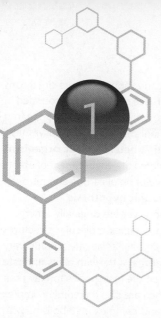

Introduction to Origins of Life Questions

This chapter introduces basic questions about the origins of life on Earth. What is life, and can we define it? What is the history of the field, and what does the modern scientific understanding have to say? What is the ribonucleic acid (RNA) world hypothesis, and why is it important? Are metabolism or genetics of greater import? These questions will be addressed in this introductory chapter. The terms 'chemical evolution' and 'prebiotic chemistry' are discussed, and a brief history of modern thinking on the origins of life is outlined.

1.1 What Is Life?

For most of human history, the origin of life was explained by a process known as spontaneous generation—an idea which has now been superseded by the rigour of modern scientific understanding, but lingers due to imprecise definitions. Aristotle wrote in his *History of Animals* that some animals did not seem to arrive from parent animals, but were created through 'spontaneous generation . . . from putrefying earth or vegetable matter'. This was an appeal to an unobserved cause, but a cause, nonetheless. This notion of spontaneous generation was passed from the Greeks to the Romans and to Christian Europe into the Renaissance. Cultures beyond Europe also asked these questions—they are after all quite fundamental.

Notions of spontaneous generation were first cohesively challenged by the experiments of the Italian scientist Francesco Redi, who demonstrated that maggots only appeared on meat which had been in contact with flies, suggesting that the flies had planted some sort of 'seed'. It was not until the 19th century that this incarnation of the idea of spontaneous generation would be laid to rest, when Pasteur proved that properly sterilized organic material did not give rise to new living organisms, including microbes.

After Pasteur's refutation of spontaneous generation in 1859, the scientific community was left with a puzzle. Evidently extant life could be explained by an evolutionary paradigm of descent with modification explained by **Darwinian evolution**, but the initial generation of life, i.e. its origins, remained inexplicable. In the early 20th century, two scientists, Oparin in Russia and Haldane in Great Britain, independently proposed origin-of-life models positing the self-organization of simple organisms from environmentally supplied organic compounds. These notions were grounded in the logic that although primary producers, e.g. photosynthetic organisms appeared to be the base of all known food chains, the mechanisms behind photosynthesis are still fairly complex. It was conceptually simpler that **heterotrophic organisms** formed first. These primordial heterotrophs then depleted the environment of abiotically supplied foodstuffs, spurring the need for the evolution of carbon and nitrogen fixation. These ideas are now known as the '**heterotrophic hypothesis**'.

Current hypotheses suggest that whenever and however life originally emerged, it did so progressively and sequentially, building up complex molecular structures and sophisticated behaviours along the way. A fly, and its larvae, are already very complex organisms; it would be silly now to consider that a fly, with all its intricate organs and evolutionary adaptations, could have been the first species to crawl out of a 'primordial soup'. The simplest bacteria or archaea are also incredibly complex structures, typically containing thousands of genes and complex metabolisms with intricate control mechanisms. It remains challenging to understand how something as seemingly simple as a single-celled organism arose. It is thought that the first cells, sometimes referred to as **protocells**, must have been much simpler than even the simplest modern prokaryote.

Can we define life?

Science progresses by the classification of phenomena and codification of the relationships among them. Following the scientific method, there must be a point where the line between life, geology, chemistry, and physics blurs. The emergence of life likely wasn't a sharp transition between something 'definitely not alive' and something 'definitely alive', but a progression of gradual and incremental changes, eventually leading to more life-like behaviours and structures over time. In fact, there is still no generally accepted definition of life among scientists even today.

While adopting a strict definition of life may help arbitrarily pinpoint an exact transition where chemistry stops and biology begins, such notions may not help us understand how chemistry made such a transition possible in the first place. Instead of trying to define life, a more fruitful exercise would be to simply *describe what life does and how it does it*. Perhaps we can identify which 'life-like' behaviours we should be looking for or trying to understand, if our goal is to understand how to make life in the lab 'from scratch', see Figure 1.1.

A likely solution to this challenge is that we need to understand chemistry that affords chemical systems capable of Darwinian evolution. This is the basis of the National Aeronautics and Space Administration (NASA)'s working definition of life, 'Life is a self-sustaining chemical system capable of Darwinian evolution', i.e. we should be looking for and trying to prepare molecular systems that can undergo evolution by natural selection.

Darwinian evolution requires *descent with modification*: the ability of a molecule or system of molecules to replicate themselves so that useful adaptations can be passed on, but in such a way that offspring are not always perfect copies of their parents, thus having the potential to develop new advantageous traits of their own. Natural selection also requires populations of individuals to act upon, wherein some individual members of a population possess traits that are more beneficial, affording them

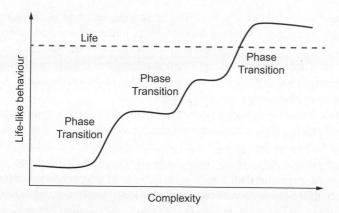

Figure 1.1 Schematic representation of the complexifaction of chemical systems over time. Note the evolution of these systems as represented here undergo sudden jumps (i.e. phase transitions) in complexity or 'life-like behaviours'. Where we draw the line for the transition from non-life to life may be arbitrary.

greater rates of reproduction at the expense of less adapted members. Importantly, to enable selection the replicating population must generate numerous variants as part of its fundamental method of self-reproduction.

This working definition of life may sound simple, but how exactly can a population of synthetic cells be created from scratch from a self-replicating system of molecules via simple chemistry? Some might say that by focusing too much on the end goal of achieving Darwinian evolution, we are, to use a modified metaphor, missing the trees by seeing only the forest. After all, Darwinian evolution is only made possible by having cells with complex underlying metabolism that can support the chemical construction of all of their molecular parts to afford replication in the first place.

Did metabolism or genetics come first?

Many scientists think metabolism is as important for the origins of life as Darwinian evolution. The way a cell does chemistry and the way humans have learned to do chemistry throughout the centuries are two very different approaches even though both must be based on the same laws of nature. While doing laboratory research, chemists often attempt to isolate the processes which affect reactions, under carefully controlled conditions to avoid producing unwanted side products. Once a reaction is finished, chemists tend to assume the reaction has come to equilibrium, i.e. optimally produced the product of interest, and nothing more interesting in terms of new chemical transformations is expected to happen. Chemists isolate, purify, and go on to the next step until they have made their desired final product.

Metabolism, on the other hand, follows fundamentally different chemical synthetic logic. The cell is a microscopic reaction vessel—a tiny 'flask'. Despite its small size (and possibly because of it), a cell can carry out thousands of different reactions simultaneously, in a coordinated and highly regulated way, connected through positive and negative feedback mechanisms so that the cell is able to manufacture what it requires responsively. These feedback mechanisms allow the cell to behave as though it 'knows what it's doing', responding to the environment to optimize its chances of success and survival, ultimately amplifying its own reproduction while relying on just a few basic inputs from the environment. These metabolic reactions almost never reach equilibrium; in fact, if equilibrium in a cell were ever reached, then one might say that cell is dead. It is interesting to think that living metabolism has been going on uninterrupted since life's origins without *ever* having come to equilibrium. If equilibrium had been attained, we wouldn't be here!

Whether Darwinian evolution or chemical reaction networks resembling metabolism should be the feature of greater importance in understanding life's emergence has been framed as the 'genetics first' versus 'metabolism first' debate. These two schools of thought still exist, but there is a growing appreciation that neither feature is likely more important than the other (or other as of yet unexplored phenomena), but that both are crucial to understanding what life is, and how it originated.

Understanding the origin of life likely requires collaboration between many scientific disciplines. While the challenge of elucidating life's origins is fundamentally a chemical problem, solving it will require an understanding of the *geo*chemical processes that preceded it, as well as the *bio*chemistry that came after. Chapter 2 will introduce important geological and geochemical background information relevant to the origins of life, and Chapters 3 and 4 will provide an overview of relevant biochemistry and a brief history of life. For the rest of this chapter, we will conclude with a discussion of the ribonucleic acid (RNA) world hypothesis and the definition of prebiotic chemistry.

1.2 The RNA World Hypothesis

What is the RNA world hypothesis?

Although we don't have a complete scientific theory of life, we do have some ideas about the problems life must have overcome before the emergence of the **last universal common ancestor (LUCA)**, the ancient organism or organisms from which all life descends. This collection of problems is largely embodied in the **RNA world hypothesis**. One difficulty in explaining life's origins is the fact that deoxyribonucleic acid (DNA) is required to make proteins, while proteins are required to make DNA. (These details will be discussed more in Chapter 3.) This situation presents a chicken-and-egg paradox—did DNA or proteins evolve first? The RNA world hypothesis resolves this dilemma by positing that there was an earlier evolutionary stage in which **RNA** played both the roles of a genetic polymer like DNA and catalytic polymers like proteins, see Figure 1.2.

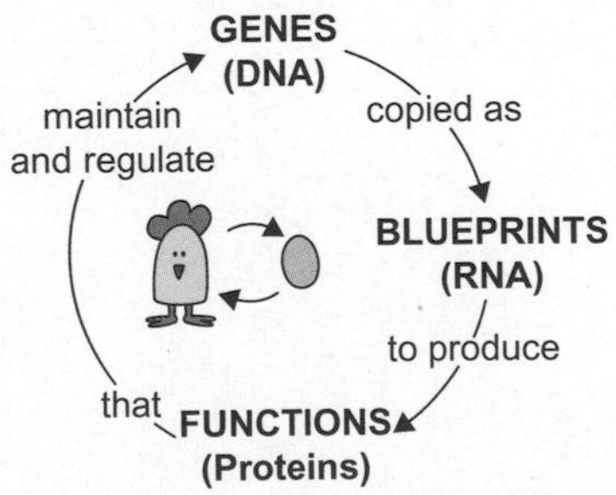

Figure 1.2 An illustration of one of the fundamental questions for the origins of life. DNA is needed to make messenger ribonucleic acid (mRNA) which is needed to make proteins, but then proteins are needed to synthesize DNA. This fact presents a chicken-and-egg dilemma about which came first, and also, how could such a complex mechanism arise in the first place.

The ribosome is a ribozyme

One of the biggest pieces of evidence in favour of the RNA world hypothesis is the fact that the **ribosome**, the ubiquitous macromolecular machine responsible for the synthesis of proteins, is mostly composed of RNA, and the catalytically active site where protein synthesis occurs is entirely based on RNA. Rather than an enzyme, the ribosome is a **ribozyme**, i.e. a catalytic molecule composed (mostly) of RNA rather than an amino acid polymer. The ribosome is often referred to as a *molecular fossil*, an evolutionary relic that provides evidence that early life evolved from primitive cells that relied more heavily on RNA for carrying out both genetic and metabolic functions. Besides the ribosome, there are a number of other ribozymes in extant cells that carry out critical metabolic tasks. Several cofactors like nicotinamide adenine dinucleotide (NADH) and flavin adenine dinucleotide (FAD), and adenosine triphosphate (ATP) itself, are also based on ribonucleotides. Another piece of evidence for the RNA world hypothesis is the fact that ribozymes can be artificially selected for in the lab using an experimental procedure that models Darwinian evolution.

Was RNA the original biopolymer?

While there is general consensus that an RNA world existed sometime prior to the appearance of LUCA, there is still considerable scientific debate as to whether RNA was also the first biopolymer to appear on Earth through **abiotic** geochemistry. It is possible that the original biopolymer was completely different, and only later evolved to generate RNA after primitive life was already established. For example, amyloid-like **peptides** have been proposed as one of the first self-replicating polymers to arise abiotically. Many studies have been conducted aiming to demonstrate the synthesis of RNA and its constituent molecular components using nonenzymatic chemistry consistent with possible early Earth geochemistry. Significant progress has been made, but much work remains to be done. The synthesis of RNA as well as peptides will be the subject of later chapters.

Can life be based on chemistry different from our own?

Since we don't yet have a complete scientific theory of life, we don't know if life can be based on different types of chemistry besides what we find in contemporary terrestrial biology. There might be forms of life that are so different from our own that we would have trouble recognizing them at all. There could be organisms somewhere (on Earth or elsewhere) that use chemistry similar to ours, except its evolution converged on a different set of **amino acids** or a different genetic polymer. We presently don't have a sufficient scientific understanding of life or the chemistry of complex systems generally to make these types of predictions with great confidence.

1.3 Prebiotic Chemistry and Chemical Evolution

In researching the origin of life, scientists often appeal to ideas of prebiotic chemistry and chemical evolution. The meanings of these terms and how they relate to more traditional ideas of chemistry are discussed.

Chemical evolution versus Darwinian evolution

The term chemical evolution was first coined in the 1950s, roughly when the modern era of research into the origins of life began. **Chemical evolution** generally refers to the physicochemical processes that converted simple organic and inorganic feedstock compounds into more and more complex sets of molecules, a subset of which at some point became capable of replicating themselves. (Sometimes the term chemical evolution is also used to refer to and include the processes of **nucleosynthesis** involving stars that produced the elements.) Once a set of molecules emerged that was capable of undergoing self-sustaining replication within a cell-like compartment, then the transition from chemical evolution to Darwinian evolution would have occurred.

Prebiotic chemistry versus 'traditional' chemistry

Prebiotic chemistry seeks to understand how compounds (typically organic ones) on early Earth during the era of chemical evolution could have been synthesized by the natural geochemical environment. As such, prebiotic chemistry is subject to a number of constraints in terms of what reagents and conditions are considered to be consistent with early Earth geochemistry, i.e. what is *prebiotically plausible*. For example, (almost all) organic solvents should be avoided since they are unlikely to be as abundant as water, and reagents which could not exist robustly in the presence of water should be omitted, e.g. butyl lithium. Elements with extremely low abundance in the **crust**, like the precious metals palladium and platinum, which are common catalysts in the laboratory, are not considered plausible, except perhaps by appealing to highly specific scenarios, which would then also have to be rare. Plausibility ultimately depends on the specific, local geochemical (or astrochemical, e.g. synthesis on meteorite parent bodies) situations that are being argued for and investigated. There are no strictly agreed upon universal metrics that can provide an objective standard for deciding what is or isn't prebiotically plausible—plausibility at the end of the day is highly context specific.

From stepwise synthesis to systems chemistry

For the majority of the history of prebiotic chemistry research (and organic synthesis as a whole), most reactions have been carried out using stepwise protocols starting with pure reagents, commensurate with the teachings and techniques for what might be termed traditional organic chemistry. There is a growing recognition by the prebiotic chemistry research community that geochemical environments would have been 'messy', leading to complex mixtures, and that this complex chemistry, which may be crucial for the emergence of life-like behaviours needs to be investigated experimentally. This appeal to complex chemical systems is referred to as **systems chemistry** and is presently at the forefront of prebiotic chemistry research.

In the chapters to come, we will describe the current state-of-the-field when it comes to the prebiotic synthesis of **sugars**, amino acids, **nucleotides**, their polymers, and protocells. The next few chapters will first discuss important geologic concepts like the origin of the Solar System and Earth, as well as important biochemical concepts like cellular organization and metabolism fundamental to all known organisms.

1.4 Summary

- NASA's *working* definition of life, 'Life is a self-sustaining chemical system capable of Darwinian evolution'.
- The first cells (protocells) must have been much simpler than even the simplest modern prokaryote as well as LUCA.
- The RNA world hypothesis is the idea that ancient cells were once dependent on RNA for both genetic information storage and catalysis.
- While there is general consensus that an RNA world existed prior to LUCA, whether RNA was also the first biopolymer to appear on Earth through abiotic geochemistry is still debated.
- Prebiotic chemistry seeks to understand how compounds important for the origins of life could have been synthesized abiotically in the natural geochemical environment.

1.5 Exercises

1. What do you think the fundamental attributes of life are?
2. Do you think enough is known about biology to formulate a definition of life?
3. Do you think the internet could become 'living'?
4. What do you think is special about carbon, as opposed to other elements, that might make it uniquely suited to being the central building block of life as we know it?
5. The term 'prebiotically plausible' is highly context specific. What do you think are some of the challenges when assessing whether a reagent/condition/system is prebiotically plausible?

1.6 Suggested Reading

1. Bill Mesler and H. James Cleaves II. (2015). *A Brief History of Creation: Science and the Search for the Origin of Life*. United States: W. W. Norton.
2. Iris Fry. (2000). *The Emergence of Life on Earth. A Historical and Scientific Overview*. United Kingdom: Rutgers University Press.
3. David W. Deamer and Gail R. Fleischaker. (1994). *Origins of Life—The Central Concepts*. United Kingdom: Jones and Bartlett Publishers.
4. Addy Pross. (2016). *What is Life? How Chemistry Becomes Biology*. United Kingdom: Oxford University Press.
5. Carol E. Cleland and Mark A. Bedau (eds). (2018). *The Nature of Life. Classical and Contemporary Perspectives from Philosophy and Science*. United Kingdom: Cambridge University Press.

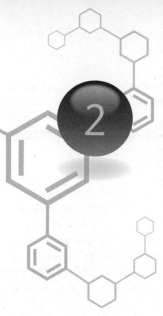

2 Origin of Earth, Its Atmosphere, and Oceans: The First Molecules

Many of the feedstocks for prebiotic aqueous-phase organic chemistry are thought to have originated in the early atmosphere or to have been delivered extraterrestrially after Earth's primary accretion. The origin of Earth and the possible composition of Earth's early atmosphere soon after its formation are discussed here. Earth's atmospheric composition has surely changed over time, as evidenced by numerous data gleaned from geochemistry. Why the consensus on this topic has changed over time from more reducing to more oxidizing atmospheres is presented. Important gas-phase reactions are highlighted, especially those driven by solar radiation (mediated by both photons and charged particles) and other energy sources, as well as chemistry resulting from extraterrestrial impacts. The composition of the early oceans, and how ocean chemistry connects to atmospheric chemistry are discussed. The chemistry of deep-sea hydrothermal vents as well as their land-based counterparts are introduced. The front of this chapter will start with important geology concepts relevant to Earth's origins and its initial atmospheric composition.

2.1 An Introduction to Geology

What is geology?

Geology is the study of Earth, its material composition, the natural processes that shape its materials, and how these processes and materials change over time. Geologists use the scientific method to understand these questions, explore how Earth formed, and what geologic properties and processes are necessary to explain presently observed Earth features.

The concept of deep time

We take it for granted that Earth is ancient, now thought to be ~4.56 billion years old. Just like deep space conveys the concept of vast distances, **deep time** connotates vast timescales. Unfortunately, the deeper something is in time or space, the more difficult it is to study. It is challenging to understand the large numbers that describe deep time and deep space, and these concepts are perhaps best understood through analogies. The concept of deep time goes beyond geology and astrophysics, however, as vast timescales are required also in the biological sciences to explain how Darwinian evolution produced the diversity of species that inhabit Earth today. Geologic timescales may also be needed for life's origins, although some think given ideal conditions, life can emerge in relatively short order.

Earth's oldest rocks

A challenge in understanding the **prebiotic** chemistry that led to life's origins is that little of Earth's earliest rock record has survived intact to the present day, because **plate tectonics** causes Earth's crust to be constantly destroyed and recycled. Subduction zones occur at plate boundaries, where the denser oceanic crust is forced downward into Earth's **mantle** where it re-enters the rock cycle. While much of the continental crust tends not to be subducted, over deep time, much of the most ancient continental crust has also been recycled.

Continental or oceanic crust older than 3.5 Ga is extremely rare. **Zircons**, a type of mineral with the formula $ZrSiO_4$, are the oldest yet-discovered crustal materials of Earth. The oldest zircons are found in the remote Jack Hills of Western Australia and date back to 4.4 Ga. The chemical properties of these zircons imply that Earth had already developed a solid crust not long after its formation. Isotopic analysis of the oxygen content of ancient zircon crystals suggests the presence of liquid water by ~4.2 Ga, see Figure 2.1. (We will discuss the processes involved in Earth's formation including when and how water arrived later).

Plate tectonics and geomagnetism

The theory of plate tectonics is central to geology, unifying seemingly disparate geological principles and observations, including the rock cycle, the origin of Earth's magnetic field, and the formation of continental crust. When in deep time plate tectonics

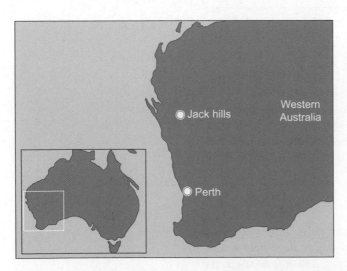
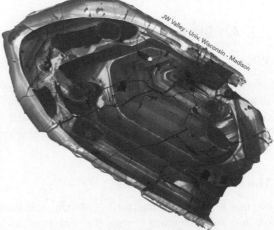

Figure 2.1 Image of one of these zircons (right) found in the remote Jack Hills of Western Australia (left). This zircon is approximately 400 microns in length.

and Earth's magnetic field originated is still a matter of debate, and the exact timing of when each started has important implications for prebiotic chemistry. Plate tectonics is responsible for more than just earthquakes and volcanoes—it is connected to the **oxidation state** of the atmosphere and helps to regulate Earth's climate. Earth's geomagnetic field also moderates the impact of energetic charged solar particles on the atmosphere. Mars is thought to once have had a magnetic field, but whose strength became dramatically diminished after its core had solidified. Whether or not a strong magnetic field is needed for the origin of life is still unknown.

Abundance of the elements in the Earth

The periodic table of elements currently contains 118 elements, but the naturally occurring elements found in Earth's crust end with uranium, which has an atomic number of 92. Of all these elements, the eight most common are oxygen, silicon, aluminium, iron, calcium, magnesium, sodium, and potassium, which together account for 99 per cent of Earth's crust, see Figure 2.2 and Table 2.1. Carbon, upon which all known life is based, does not even account for 0.1 per cent. If we consider Earth as a whole including crust, mantle, and core—iron is the most abundant element. If we look at our entire Solar System, hydrogen is by far the most abundant element due to the enormous mass of the Sun.

   In understanding the distribution of elements in Earth's layers, in particular, the crust and atmosphere where the most important prebiotic chemistry likely took place, we first need to understand how our Solar System formed. These considerations will be the topic of the next few sections.

2.2 Formation of Earth and Its Geological History

Formation of the Solar System

Billions of years of nucleosynthesis within stars afforded the elements beyond hydrogen, helium, and lithium we have in the Solar System today, of which the Sun and planets are composed. Many of these elements are also necessary for life. The

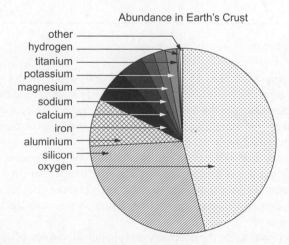

Figure 2.2 Abundance of the elements in Earth's crust.

Table 2.1 Abundances of the 30 most abundant elements in Earth's crust.

Element		
Z	Name	ppm
8	oxygen	461,000 (46.1%)
14	silicon	282,000 (28.2%)
13	aluminium	82,300 (8.23%)
26	iron	56,300 (5.63%)
20	calcium	41,500 (4.15%)
11	sodium	23,600 (2.36%)
12	magnesium	23,300 (2.33%)
19	potassium	20,900 (2.09%)
22	titanium	5,650 (0.565%)
1	hydrogen	1,400 (0.14%)
15	phosphorus	1,050 (0.105%)
25	manganese	950 (0.095%)
9	fluorine	585 (0.0585%)
56	barium	425 (0.0425%)
38	strontium	370 (0.037%)
16	sulfur	350 (0.035%)
6	carbon	200 (0.02%)
40	zirconium	165 (0.0165%)
17	chlorine	145 (0.0145%)
23	vanadium	120 (0.012%)
24	chromium	102 (0.0102%)
37	rubidium	90 (0.009%)
28	nickel	84 (0.0084%)
30	zinc	70 (0.007%)
58	cerium	66.5 (0.00665%)
29	copper	60 (0.006%)
60	neodymium	41.5 (0.00415%)
57	lanthanum	39 (0.0039%)
39	yttrium	33 (0.0033%)
7	nitrogen	19 (0.0019%)

Solar System formed between 4.5 and 4.6 billion years ago from the gravitational collapse of a section of a molecular cloud that was made up of gas and dust, forming a **protoplanetary disc** with the Sun at its centre. The original molecular cloud from which our Solar System formed was probably similar to the Orion Nebula, which can be seen with the naked eye in the night sky.

Scientific advances in astronomy, particularly over the past couple of decades, have allowed us to observe protoplanetary discs around other stars that help us understand how our own Solar System formed, but each of these discs are constructed from their own unique elemental distributions based on the molecular clouds from which they were born.

To gain direct insight into the chemistry and elemental makeup of our own Solar System, scientists study interplanetary dust and the **meteorites** that fall to Earth.

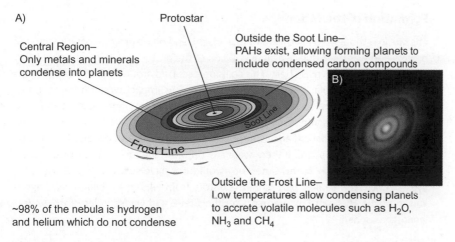

A)

Protostar

Central Region–
Only metals and minerals
condense into planets

Outside the Soot Line–
PAHs exist, allowing forming planets to
include condensed carbon compounds

B)

Soot Line

Frost Line

Outside the Frost Line–
Low temperatures allow condensing planets
to accrete volatile molecules such as H_2O,
NH_3 and CH_4

~98% of the nebula is hydrogen
and helium which do not condense

Figure 2.3 A) Illustration of a protoplanetary disc and its features. **B)** Actual image of HL Tau, a young star, and its protoplanetary disc.

Most meteorites are thought to originate from the asteroid belt that lies between Mars and Jupiter, while a small fraction have come from Mars or the Moon. The celestial bodies in the asteroid belt are thought to be the ancient remnants of the Solar System's protoplanetary disc and therefore can provide clues as to what materials the Earth and Solar System were formed from, and the dynamic processes involved. CI **chondrite** meteorites, which are considered to be 'pristine' samples of the raw material of the early Solar System, have been especially useful in this regard. When combined with spectroscopic analysis of the Sun to determine its elemental makeup, we can quantify the elemental abundances of the Solar System as a whole, setting a foundation for understanding the Earth's chemical makeup, particularly for the crust and atmosphere where prebiotic chemistry likely occurred, see Figures 2.3A and B.

Accretion of the Earth

As the Solar System was forming from the original molecular cloud, radiation from the Sun led to the formation of distinct regions of the protoplanetary disc. Hotter temperatures closer to the Sun permitted the condensation of relatively non-volatile materials such as metals and silicate minerals. Farther from the Sun, at cooler temperatures, more volatile materials condensed and accumulated. This dynamic explains why the four inner planets, including Earth, are rocky whereas the four outer planets are gaseous.

The Earth is thought to have originally formed from dust-sized mineral grains held together initially by electrostatic attractions that formed clumps of matter, until the clumps became big enough that gravitational attraction took hold. The clumps grew larger over millions of years until eventually reaching the size of small bodies known as planetesimals in a process called **accretion**. These planetesimals continued to collide with each other in violent events until only a few larger planets were left.

Formation of the Moon

The Moon is thought to have formed towards the end of the protoplanetary accretion process and was the result of a giant collision of the proto-Earth with a body roughly the size of Mars known as Theia. This collision occurred around 4.5 billion years ago, not long after the Solar System started to form. This impact would have been tremendously violent, releasing enormous amounts of energy and heat, so much so that whatever Earth's atmosphere was at that point would have been lost, and the entire surface and most of Earth's interior melted into a **magma ocean** blanketed by a rock-vapour atmosphere. If there were any primitive life forms or advanced stages of prebiotic chemistry at this time, they would not have survived this catastrophic event. In fact, the earliest part of Earth's history from ~4.5 to 4.0 billion years ago is known as the **Hadean** Eon, named after the Greek god Hades, ruler of the underworld, that is, hell.

The magma ocean

In Earth's magma ocean state, the heaviest elements (e.g. iron and nickel) sank to form the core, bringing along with them the more minor siderophile ('iron loving') elements. Conversely, lighter elements (e.g. oxygen, aluminium, and silicon) migrated upwards to form the mantle and crust, bringing along with them the chalcophile ('sulfide-ore loving') and lithophile ('rock loving') elements. This process is known as **differentiation**. There are, however, an anomalously high amount of siderophile elements in the Earth's crust, materials which were likely delivered to the Earth via meteorite bombardment after the magma ocean had cooled and formed a solid crust. The extra mass added after differentiation is known as the **late veneer**. Differentiation also had a huge impact on Earth's early atmosphere, a topic which will be discussed more later, see Figures 2.4A and B.

The late-heavy bombardment

Even after the Moon-forming impact, the early Solar System was still a turbulent place, with the Earth and other planets being bombarded by impacts at a much higher rate than Earth experiences today. Evidence for a high early impact rate can be seen from the numerous large craters on the Moon (particularly the black Mare) and the innermost planet Mercury. Some of these impacts could have been large enough to boil off the oceans or partially melt the crust, perhaps effectively sterilizing the young planet of any nascent primitive life. As the Solar System matured, however, the frequency of these large, life-destroying impacts decreased. Detailed analysis of the Moon's craters indicates that there was still a high frequency of impacts between 4.1 to 3.8 Ga, a time known as the **late-heavy bombardment**.

While some suppose that this violent era would have been too intense to allow life to emerge, others suspect that these meteorite impacts were crucial for prebiotic chemistry, helping to generate the organic feedstocks needed for life to get started. The late-heavy bombardment straddles the end of the Hadean and the beginning of the **Archean** eons. It also should be noted that the earliest direct geological evidence for life in the rock record is ~3.5 Ga. This observation could be taken as evidence that the violent era of heavy bombardment needed to stop before life could emerge. Alternatively, perhaps it was the geochemistry resulting from these impacts that was

A)

B)

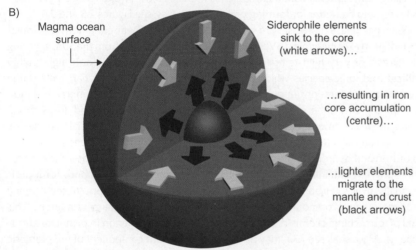

Magma ocean
surface

Siderophile elements
sink to the core
(white arrows)…

…resulting in iron
core accumulation
(centre)…

…lighter elements
migrate to the
mantle and crust
(black arrows)

Figure 2.4 A) Illustration of what the immense violence of the Moon-forming impact may have looked like leading to a magma ocean. **B**) Schematic of the process of differentiation that would have resulted when Earth existed as a magma ocean.

a prerequisite for life. An additional consideration is that the rock record older than 3.7 Ga is extremely scarce and metamorphosed, such that any possible traces of life before then has been erased by the ravages of time.

2.3 Earth's First Atmospheres: From Strongly to Weakly Reducing Mixtures

Earth's early atmosphere may have played a role in generating its first prebiotic organic molecules. The composition of the early atmosphere likely affected what organic molecules may have been abundant on Earth's early surface. In this section, we discuss Earth's first atmospheres.

The importance of atmospheric redox states

One of the main factors determining the composition of the early atmosphere is its oxidation state, which has large consequences for the types and yields of organic compounds the atmosphere can produce through various reaction mechanisms. Recall that oxidation states (also called oxidation numbers) of specific atoms in a molecule can be determined by summing up the bonds to more electronegative atoms (typically oxygen, but also sometimes nitrogen and sulfur) and subtracting the number of bonds to hydrogen. The carbon in CO_2 has four bonds to oxygen (in the form of two double bonds) and none to hydrogen and so has a +4 oxidation state (the highest oxidation state for carbon), while the carbon in CH_4 has a –4 oxidation state (the lowest oxidation state for carbon). The nitrogen atoms in N_2 both have an oxidation state of zero, since they have no bonds to either hydrogen or oxygen and are only bonded to themselves. In contrast, ammonia, NH_3, has three H atoms singly bonded to nitrogen, giving the central N atom a formal –3 oxidation state. Generally speaking, atmospheres containing gaseous molecules with more reduced oxidation states tend to be better for forming organic molecules, see Figures 2.5 and 2.6.

When gas-phase molecules are subjected to highly energetic conditions (e.g. ionizing radiation like ultraviolet (UV) light, solar energetic particles from the Sun, lightning discharges, or very high temperatures, see Table 2.2), they may form high-energy **radical** and ionic species which tend to relax to more thermodynamically stable electronic configurations upon cooling or de-excitation. Thus, the general oxidation state of the atmosphere governs whether electrons can be 'pumped' from strong bonds to weaker, more reactive ones. C–O bonds are stronger than C–H bonds due to the greater electronegativity differences between carbon and oxygen compared to carbon and hydrogen. As a result, in the presence of abundant O as is the case for atmospheres containing molecules in highly oxidized states, O–X bonds form preferentially over C–H bonds. Thus, highly oxidized atmospheres tend to inefficiently produce organic compounds even when subjected to high-energy conditions. This kind of 'quenching' chemistry by O is an important phenomenon in planetary atmospheres. As we shall see, CO_2 may have been a dominant component of the prebiotic atmosphere and thus an abundant source of carbon, but is not especially reactive.

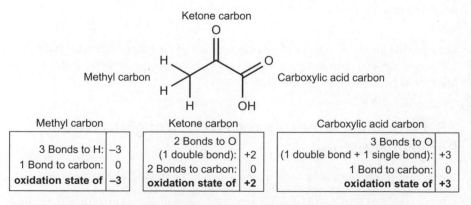

Methyl carbon		Ketone carbon		Carboxylic acid carbon	
3 Bonds to H:	–3	2 Bonds to O (1 double bond):	+2	3 Bonds to O (1 double bond + 1 single bond):	+3
1 Bond to carbon:	0	2 Bonds to carbon:	0	1 Bond to carbon:	0
oxidation state of	**–3**	**oxidation state of**	**+2**	**oxidation state of**	**+3**

Figure 2.5 Worked example of the oxidation states for each carbon in the molecule pyruvate, a ubiquitous biochemical metabolite, which is also considered to be prebiotic.

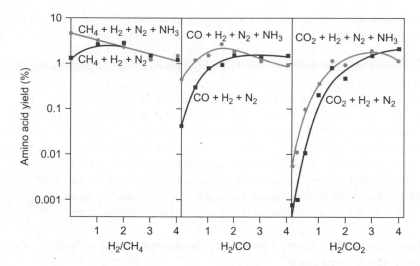

Figure 2.6 Amino acid yield obtained from spark discharge experiments as a function of gas composition. Graph reproduced from G. Schlesinger, S. L. Miller (1983). 'Prebiotic Syntheses in Atmospheres Containing CH_4, CO, and CO_2: I. Amino Acids'. *J Mol Evol* 19: 376–382.

Note: Greater ratios of CO or CO_2 to H_2 tend to result in lesser yields.

Table 2.2 Energy sources on the modern Earth. Table adapted from 'Prebiotic Chemistry on the Primitive Earth'. 2007. Oxford University Press.

Source	Energy (cal cm^{-2} yr^{-1})	Energy (J cm^{-2} yr^{-1})
Total radiation from Sun	260,000	1,090,000
Ultraviolet light <300 nm	3,400	14,000
Ultraviolet light <250 nm	563	2,360
Ultraviolet light <200 nm	41	170
Ultraviolet light <150 nm	1.7	7
Electric discharges	4.0	17
Cosmic rays	0.0015	0.006
Radioactivity (to 1.0 km)	0.8	3.0
Volcanoes	0.13	0.5
Shock waves	1.1	4.6

The highly reducing primordial atmosphere of Earth was quickly lost

Earth's current atmosphere is composed of 78 per cent nitrogen, 21 per cent oxygen and less than 1 per cent of other gases including argon, water vapour, and CO_2. It is generally thought Earth's earliest atmosphere was more reducing than at present. There are several lines of reasoning for this hypothesis. First and foremost is that the majority of the oxidants (e.g. O_2) present in the modern atmosphere are derived from biological influences. Modern plants and marine photosynthesizing organisms produce enormous quantities of O_2 that appear to have been abiologically titrated (consumed by redox recombination) by massive geological reservoirs of reductants such as iron and sulfur and thereby depleted O_2 without it being continuously replenished. While the prebiotic atmosphere likely lacked abundant O_2, could it have been even more reducing?

It was once thought that Earth's prebiotic atmosphere closely resembled those of the gas giants and was highly reducing, however, there is now a general consensus that a highly reducing terrestrial atmosphere could have only existed over brief geologic timescales prior to the Moon-forming impact. During the protoplanetary phase as Earth was accreting from planetesimals, nebular gases were still present, the most abundant of which would have been H_2. In this scenario, it is possible that planetesimals and protoplanets could have acquired a **primary atmosphere** composed predominantly of H_2—that is, highly reducing. Even if this were the case, however, the primary atmosphere would have been lost as a consequence of the extreme violence of the Moon-forming impact as well as the general removal of redox buffers such as iron to Earth's interior.

Why the oxidation state of the mantle matters for the oxidation state of the atmosphere

Following the Moon-forming impact, Earth acquired its **secondary atmosphere** based on the volatile compounds that outgassed from its interior. Recall that this immensely violent event would have resulted in a magma ocean. This magma ocean likely lasted for millions of years, as the metallic iron from the impactor eventually differentiated and sank into Earth's lower mantle and core.

Once differentiation was complete, it would have left behind a comparatively oxidized mantle, which in turn governed the oxidation state of the volatiles which were released from it during volcanic degassing. Hence, if the early atmosphere was derived from volcanic outgassing, the oxidation state of the mantle and atmosphere would have been effectively coupled. Outgassing through volcanic eruptions would have transitioned the atmosphere to a more oxidized state dominated by H_2O, N_2, and CO_2, with perhaps as much as 200 bars of the latter, with only a small amount of CO and H_2 and little methane. This atmosphere derived from outgassing dominated by water vapour and CO_2 may have lasted throughout the rest of the Hadean (until ~4 Ga). CO_2 was likely removed from the atmosphere as it is today through formation of carbonates (especially calcium and magnesium ones), but the timescale for this process is not well constrained and also depends on the onset of plate tectonics, which recycles carbonates back into CO_2 through volcanic outgassing.

Before differentiation was complete, however, residual reduced iron in near-surface environments could have effectively acted as a reductant that may have yielded initially a more reducing atmosphere with significant amounts of H_2 and CO. How long this mildly reducing atmosphere lasted would have depended on the rate of differentiation.

Transiently reducing atmospheres might have occurred

From a geologically global perspective, Earth's early atmosphere during the age of prebiotic chemistry (~4.4 to ~3.5 Ga) was likely not as reducing as the gas giants' as it was once thought to be, instead being composed mostly of water vapour, CO_2, N_2, and smaller amounts of CO and H_2 as a consequence of the mantle redox state. Nevertheless, there were likely periods during which more reduced atmospheres could have been transiently acquired locally or even globally and which could have facilitated abundant atmospheric organic molecule synthesis. For example, meteorite impacts themselves can create reducing atmospheres by virtue of their abundance of metallic iron and the kinetic energy of the impact itself. Additionally, a process known as **serpentinization** that occurs when water interacts with certain crustal minerals can afford H_2 locally. These types of processes will be discussed in more detail in the next sections.

2.4 The First Organic Molecules: Exogenous Delivery, Endogenous Production, and Impact Synthesis

There are two general mechanisms by which Earth can abiotically obtain organic molecules: (i) molecules can be synthesized elsewhere where favourable conditions exist and then transported to the Earth, that is, exogenous (or extraterrestrial) delivery, and (ii) organic molecules can be synthesized directly on Earth from available feedstocks, that is, endogenous production. There are a variety of mechanisms through which abiotic organic synthesis could have occurred endogenously through atmospheric chemistry on early Earth, that can be classified according to the energy sources involved in instigating the synthesis. These energy sources include UV light, solar energetic particles, **electric discharges** (lightning), as well as kinetic energy delivered through cometary and meteorite impacts. A third way, which is in some sense a hybrid of exogenous and endogenous production is what is known as impact synthesis, a process which can promote the reaction of molecules already present in the atmosphere as well as those brought by the impactor itself. Examples of each will be discussed later.

Extraterrestrial organic synthesis and delivery

The modern Earth receives approximately 1.4×10^7 kg of **extraterrestrial** material each year, mainly in the form of interplanetary dust particles, which are micron- to mm-sized ice and dust grains probably most relatable to the reader as the cause of common shooting stars observed on clear moonless nights, and to a much lesser extent meteorites, which are larger rocky or metallic bodies. Indeed, several large recent infalls have been recorded on film (e.g. a multi-metre sized meteor was caught on film exploding over the Russian city of Chelyabinsk in 2013), see Figure 2.7.

In addition to the small meteorite chunks which are routinely recovered around the world, Earth's surface records many large meteorite impacts in the form of meteor craters. Scientists can estimate the relative flux of such impactors over time to the Earth's surface by observation of the Moon's heavily cratered surface and infer that the mass input of ET impactors was many orders of magnitude greater early in Earth's history.

Since many types of these bodies are now known to contain significant amounts of organic material, it follows that these could have been a significant source of organics to the early Earth, especially if the early terrestrial atmosphere was not very reducing. In any event, there is some similarity in the nature of the organics produced by atmospheric processes and those detected in extraterrestrial materials.

Besides the identifiable materials suspected to have contributed to Earth's formation, various other remnants of early Solar System collisions which did not condense to form planets remain observable, namely the asteroid belt and the comets, which make up the so-called **Oort cloud** and **Kuiper belt**, see Figure 2.8.

Figure 2.7 Photograph of a meteor exploding in the atmosphere over Chelyabinsk, Russia recorded in 2013.

The asteroidal materials, which orbit the Sun between ~2.2 and 3.2 AU, are generally rockier and more metallic, while comets tend to be icier. These bodies generally obey circular though somewhat chaotic orbits, and from time-to-time they are ejected from their stable orbits and may cross the Solar System, occasionally striking planetary surfaces, including Earth.

In general, the organic content of comets is of the order of a few per cent, while that of asteroids is also of the order of 1–2 per cent, depending on the degree of maturity of the asteroid. Meteorites are thought to be fragments of asteroids generated by inter-asteroidal collisions, while the frequent shooting stars observed on Earth are thought to be remnants of material ejected by comets during their periodic Solar System wanderings, which explains why many meteor showers are predictable.

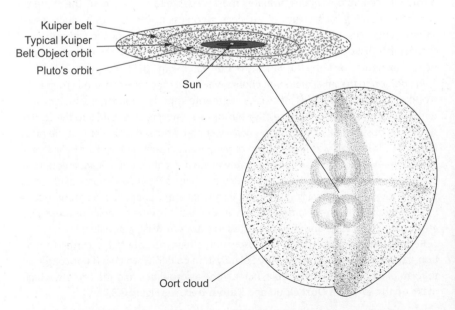

Figure 2.8 Illustration showing where the Oort cloud and Kuiper belt are in the Solar System.

Impact synthesis

While some asteroids and meteorites are particularly carbon-rich, a far more common type are native-metal rich, meaning they contain metals in the zero-oxidation state. Such bodies also likely impacted the early Earth frequently and could have promoted prebiotic organic synthesis by another mechanism—reducing the early atmosphere itself—which would have enabled organic syntheses via lightning and UV light. In other words, as such masses of metal moved through the early atmosphere at high velocity, they would have created locally (and perhaps even globally) reduced environments conducive to organic synthesis, see Figure 2.9.

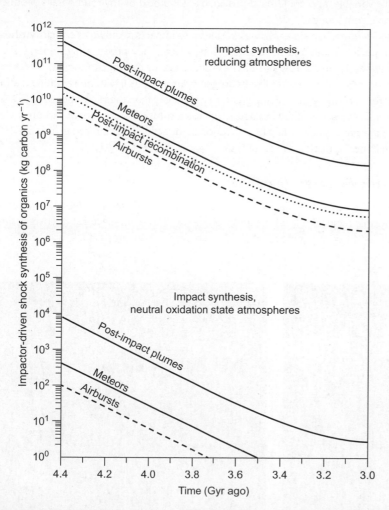

Figure 2.9 Synthesis of organics in the terrestrial atmosphere on early Earth over time as driven by shocks produced by impacts. The upper curves represent reducing atmospheres containing CH_4 + (N_2 or NH_3) + H_2O and the lower curves CO_2 + N_2 + H_2O neutral atmospheres. Dashed lines signify upper bounds, and the dotted line is an estimate that is not well constrained. Graph reproduced from C. Chyba, C. Sagan. (1992). 'Endogenous Production, Exogenous Delivery and Impact-Shock Synthesis of Organic Molecules: An Inventory for the Origins of Life'. *Nature* 355: 125–132.

We know that the early Earth must have undergone a period of frequent impacts after cooling of the magma ocean thanks to the late veneer discussed earlier. Upon impact, the metallic iron together with the intense heat could have served to generate a more reduced atmosphere, producing hydrogen gas H_2 and carbon monoxide CO, and for larger impactors potentially even methane CH_4 and ammonia NH_3. Recently, it has been postulated that sufficiently large impacts could have induced transient (up to 100 million year) periods of extremely reducing atmospheric conditions during Earth's earliest history, affording abundant organic molecules. On the other hand, the largest impactors would have been energetic enough to boil off the entire oceans, likely sterilizing any life or life-like processes that managed to gain a foothold. The best guess is that the period during which life formed on Earth was highly variable, and could have provided a good deal of environmentally produced organic compounds.

Another mechanism of synthesis relevant to impacts is known as **shock synthesis**. As impactors enter and traverse the atmosphere, they produce high-energy shock waves, see Figures 2.10A and B. These shock waves can initiate the gas-phase production of small molecules like hydrogen cyanide (HCN), with more reduced atmospheres containing CH_4 having greater productivity than those composed of CO_2, N_2, and H_2O. Moreover, if the impactor contains a substantial amount of carbon, friction with the atmosphere will cause it to ionize directly, affording molecules like HCN even in non-reducing atmospheres if N_2 gas is present.

Electric discharges (lightning)

One of the earliest mechanisms investigated for the endogenous prebiotic production of organic molecules from the atmosphere was the use of electrical discharges simulating lightning, experiments of which employed a reducing gaseous mixture that was

A)

B)

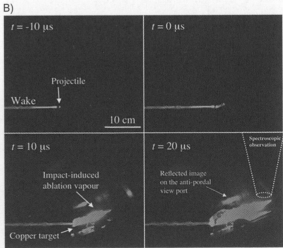

Figure 2.10 A) A high-velocity impact gun that simulates meteorite impacts. **B)** High-speed photographs taken as a polycarbonate projectile impacts a copper target under an N_2 atmosphere. Note the plume of gas that forms in the bottom-right image in which molecules like HCN can be detected spectroscopically. Images reproduced from N. Kawai et al. (2015). 'Stress Wave and Damage Propagation in Transparent Materials Subjected to Hypervelocity Impact'. *Procedia Eng 103*: 287–293; K. Kurosawa et al. (2013). 'Hydrogen Cyanide Production due to Mid-Size Impacts in a Redox-Neutral N_2-Rich Atmosphere'. *Orig Life Evol Biosph 43*: 221–245.

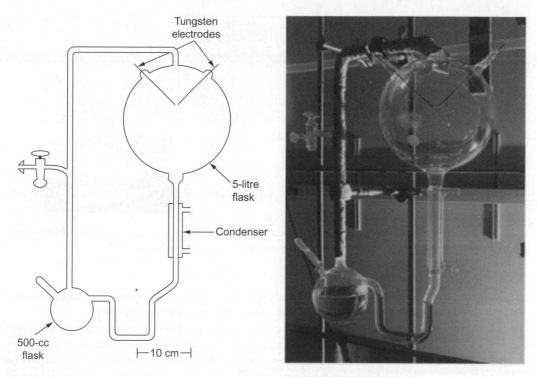

Figure 2.11 The Miller–Urey Experiment apparatus. The lower flask contained water, simulating the primitive oceans, while the upper flask was embedded with two tungsten electrodes which allowed the passage of an electric arc between them, simulating lightning. The water in the lower flask was boiled using a hot plate, and the steam circulated into the electric discharge chamber and was then condensed back into the lower flask, bringing along with it small organic molecules generated in the gas phase by the electric discharge.

supposed at the time to mimic the early Earth atmosphere. In 1953, Stanley Miller, a young graduate student inspired by his advisor Nobel Prize winning chemist Harold Urey's thoughts on the nature of the primitive atmosphere, conducted a ground-breaking experiment demonstrating the plausibility of organic compound synthesis from the action of electric discharges simulating lightning on reduced gas mixtures composed of H_2, CH_4, NH_3, and water vapour (Figure 2.11). Electric discharges, such as lightning, are able to affect chemistry by breaking bonds through a combination of high temperature, electron generation (ionization), and UV/X-ray emission. A cloud of radical species is thus generated which recombines distal from the discharge. Miller was able to show that among the most abundant initial products were simple species like HCN and formaldehyde (HCHO), along with other nitriles and carbonyl-containing species and even amino acids. This experiment will be discussed more in Chapter 6.

UV light and organic hazes

While lightning strikes are quite common on Earth (with ~44 strikes per second on the modern Earth, each discharging the order of a gigawatt of power), UV light was likely a much more important source of energetic processing in the primitive atmosphere, though its effects are attenuated by passage through the atmosphere. While the atmosphere is much denser in the lowest level (the troposphere) where molecular collisions are favoured, the penetration of chemistry-inducing high-energy

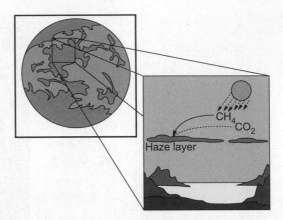

Figure 2.12 Depiction of a possible early Earth haze which could have formed from abundant atmospheric methane and solar UV radiation.

radiation is also much less. The shortest wavelengths that could have reached the surface of early Earth were ~200 nm (as will be discussed).

NH_3 and CH_4, which are reactive molecules in atmospheric organic synthesis, are also UV reactive, and thus have short half-lives on geologic timescales. They are prone to photolysis, and their existence in the prebiotic atmosphere would have been short lived without a continual resupply. NH_3 is particularly vulnerable, since it undergoes photolysis at wavelengths <215 nm, which were likely able to reach the Earth's surface. If Earth's atmosphere at some point contained abundant enough CH_4, then UV photolysis in the upper atmosphere would have triggered organic **haze** formation, not unlike that observed on Saturn's moon Titan, see Figure 2.12.

These haze particles and their constituent organics could have been deposited at the surface, affording further aqueous-phase chemistry. Meanwhile, the haze itself could have acted as a shield against UV light reaching the surface, although the effectiveness depends on the haze particle sizes, morphologies, and concentrations.

Solar energetic particles

The Sun also emits a constant stream of high-energy ionizing charged particles known as the **solar wind**. These particles interact with the upper atmosphere and Earth's magnetic field to produce the fantastic auroral light shows at the poles, that is, the northern and southern lights. A much higher flux of charged particles can emerge from occasional **coronal mass ejections**, which are powerful enough to disrupt electrical equipment on Earth's surface, see Figure 2.13.

It has been suggested that the young Sun produced more of these solar energetic particles in addition to more frequent and intense coronal mass ejection events. It is also suspected that Earth's magnetic field may have been weaker during this time and so less shielded from these high-energy events. The resulting greater flux of high-energy particles may have affected the composition of the atmosphere, which would have had implications for prebiotic chemistry occurring at the surface. Modelling studies suggest that these types of radiation may have contributed to the chemistry of both the upper and lower atmosphere of the early Earth, potentially affording the production of HCN and NO_x species in the lower atmosphere.

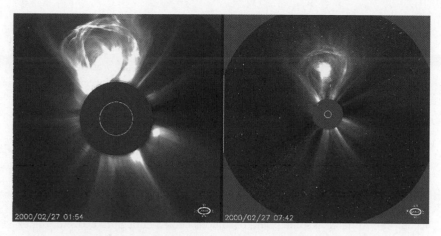

Figure 2.13 Photograph of a coronal mass ejection from the Sun.

The young Sun also put out more UV radiation

While the young Sun was likely fainter overall, it nevertheless also used to output more radiation in UV wavelengths in comparison to today. Given the lack of an ozone layer and the assumption of an atmosphere dominated by CO_2 and N_2 with little-to-no reducing gases, the surface of the primitive Earth would have been exposed to a far greater flux of UV light in comparison to today, particularly in the ~200 to 300 nm region, see Figure 2.14.

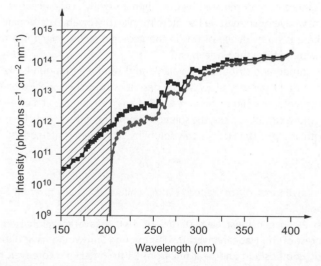

Figure 2.14 Graph of the solar spectrum (upper curve, squares) overlaid with the spectrum reaching the Earth's surface after passing through a likely prebiotic atmosphere dominated by CO_2 and N_2 (lower curve, circles). The shaded region represents absorption primarily due to CO_2. Graph reproduced from S. Ranjan, D. D. Sasselov. (2016). 'Influence of the UV Environment on the Synthesis of Prebiotic Molecules'. *Astrobiol 16*: 68–88.

Note: Wavelengths > ~200 nm would have reached the Earth's surface.

These UV wavelengths have important implications for both atmospheric and aqueous prebiotic chemistry, as these photons are energetic enough to affect the electronic structures of organic molecules and metal complexes alike, affording the production of larger molecular weight compounds in some cases, as well as decomposition in others. In this sense, UV light is a double-edged sword for prebiotic chemistry, affording both the production and destruction of organic molecules.

2.5 Oceans and Submarine Hydrothermal Vents

The earliest stages of Earth's surface evolution, including the conditions of the first oceans, are difficult to constrain. When Earth first formed, and immediately after the Moon-forming impact, Earth's surface was a magma ocean enveloped by a rock-vapour atmosphere. At some point, conditions cooled enough to allow for the existence of surface water. The presence of liquid water could have occurred within 100 Ma of the Moon-forming impact or could have taken somewhat longer. During this period, the Earth went from being a body with no ocean coverage to one with essentially 100 per cent ocean coverage. The appearance and growth of the first land masses are still topics of debate, as discussed below.

Formation of the early ocean after the Moon-forming impact

The first oceans were likely salty from the start, because the rock-vapour atmosphere resulting from the Moon-forming impact would have included the relatively volatile sodium chloride, which would have remained in the atmosphere for a comparably long amount of time before condensing on the surface. Other ions possibly dissolved in relatively high concentrations were potassium, magnesium, and iron (Fe^{2+}). If the early atmosphere were dominated by CO_2, then a slightly acidic ocean is also likely, and there may have been massive but more mobile (dissolvable) carbonate deposits, although there is no evidence of this in the geologic record, since few rocks have survived unaltered from the Hadean Eon.

The environmental distribution of volatile and soluble compounds like CO_2 and NH_3 are governed by Henry's Law, which describes how such species partition between a given volume of liquid and pressure of an overlying gas at a given temperature, and in more complex cases, the solution chemistry of the liquid phase, including ionic strength and pH. The Henry's Law solubility coefficient is defined as:

$$H^{cp} = c_a/p$$

where H^{cp} is given in mol/(m^3·Pa) or M/atm, while c_a is usually given in M and p in atm, see Table 2.3.

Although the ocean must have started off hot, its average global temperature during the rest of the Hadean given the faint young Sun would have depended on atmospheric composition and heat flux from Earth's interior. Moreover, higher atmospheric pressures raise the boiling point of water. Thus, the average temperature of the prebiotic ocean is difficult to constrain over the course of the Hadean. Furthermore, the length of a day (e.g. a full Earth revolution) would have been much shorter, ~10 h, and the Moon much closer, resulting in more frequent and greater amplitude tides, perhaps averaging as high as 20 metres.

Table 2.3 Henry's law constants for some common atmospheric gases.

Gas	Henry's Law Constant (M atm^{-1})
He	3.7×10^{-4}
Ne	4.5×10^{-4}
N_2	6.1×10^{-4}
H_2	7.8×10^{-4}
CO	9.5×10^{-4}
O_2	1.3×10^{-3}
Ar	1.4×10^{-3}
CO_2	3.4×10^{-2}

Organic chemistry in the open ocean

The Hadean ocean was probably a reservoir for some of the organic and inorganic molecules produced by the various atmospheric and exogenous delivery processes discussed above, potentially accumulating steady-state concentrations of more abundant species. For example, a steady-state concentration for HCN of 2 μM for an ocean at pH 8 and 0 °C has been estimated, but these sorts of calculations are difficult to constrain. While the early ocean certainly fostered a complex suite of organic molecules and reactions, it has often been suggested that life could not have originated in the open ocean because the concentration of organic molecules would have been too dilute to form molecules thought to be crucial, like genetic and catalytic polymers. Nevertheless, the chemistry at the oceans' surface probably would have been quite distinct from that occurring in its depths, being exposed to the atmosphere and UV light, and may have even accumulated immiscible hydrocarbon 'slicks' analogous to oil on water.

Submarine hydrothermal vents

The majority of prebiotic chemistry research when it comes to ocean-based scenarios has focused on **submarine hydrothermal vents**. Submarine hydrothermal vents, first discovered in the 1970s, immediately sparked scientific curiosity as a potential geochemical locale for the birthplace of life. Such vents presently teem with exotic lifeforms and occur in the vicinity of active tectonic areas like spreading centres or subduction zones, with the former being posited as most relevant to prebiotic chemistry.

There are two general types of submarine hydrothermal vents, black smokers and white smokers, each characterized by ornate mineral assemblages known as chimneys. Black smokers form directly above spreading centres, while white smokers are located farther from the primary heat source. White smoker chimneys may have been more common during the age of prebiotic chemistry, since a significant portion of the sulfide chemistry of modern black smokers depends on the biological generation of abundant seawater sulfate, see Figure 2.15. White smokers, on the other hand, are characterized by alkaline, mineral-rich fluids containing reducing gases including H_2. Upon coming into contact with the cooler, slightly acidic seawater of the Hadean seafloor, the dissolved minerals would have precipitated forming white smoker chimney structures similar to those observed today.

The H_2 produced by the formation of white smoker fluid is formed through the reaction between water and the underlying rock in a process known as *serpentinization*. Seawater percolates through cracks and fissures of the oceanic crust, which is rich in ferromagnesian minerals like olivine $((Mg,Fe)_2SiO_4)$. At these depths, the pressures and temperatures are great enough for the ferromagnesian minerals to reduce water, forming H_2 and oxidizing the minerals in the process. This heated fluid then flows back towards the surface where it emerges from the vents. It has been proposed that this H_2 would have been capable of reducing dissolved CO_2 in Hadean seawater in reactions catalysed by mineral catalysts within the chimneys, producing organic molecules. These newly formed organics could have been prevented from diluting into the bulk ocean by accumulation within the chimney pores. Hence, submarine hydrothermal vents like white smokers have come under intense investigation and scrutiny as candidate locations for life's emergence.

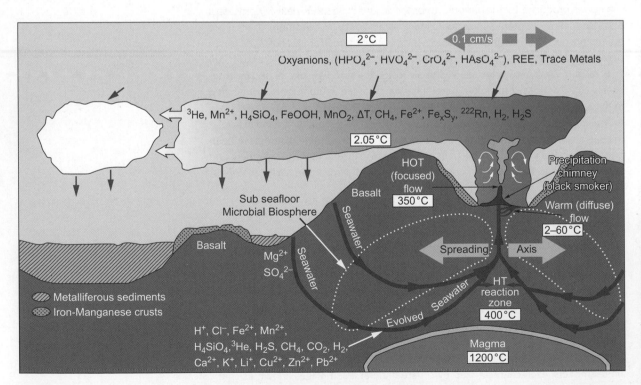

Figure 2.15 Diagram of the dynamics of a submarine hydrothermal system resulting in a black smoker chimney. Diagram reproduced from the National Oceanic and Atmospheric Administration.

2.6 Land and Hydrothermal Fields

Did the early Earth have dry land? To answer this question, we need to know two things: (i) how deep were the oceans and (ii) how high were Earth's surface features? If the surface of the early Earth were incredibly smooth, then a shallow ocean would have sufficed to cover the entire surface. On the other hand, if the early oceans were especially deep, then only very large surface features could have been exposed above sea level. Moreover, there is no reason to assume the volume of the ocean has been constant, and its mean depth could have oscillated on geologic timescales during the Hadean.

What were the first landmasses like?

The magma ocean following the Moon-forming impact likely left Earth's surface relatively smooth, and the earliest ocean that formed after the crust had cooled enough probably was still shallow, since degassing of water from the mantle takes time. Even though continental crust may not have existed during the time of life's origins before 3.5 Ga, there were nonetheless likely pockets of dry land in the form of volcanic islands caused by **geologic hotspots** via a process that does not require plate tectonics. Hotspots occur where a plume of especially hot mantle rises upwards and erupts through the crust resulting in volcanism that, over time, can form islands that rise above the surface of the ocean, see Figure 2.16. The Hawaiian Islands are a famous example of hotspot volcanism, and a similar process that formed these islands is considered likely to have produced the first dry land that protruded above the surface of the oceans on the primordial Earth.

If plate tectonics started early in Earth's history, then it is possible that, by the mid-Hadean, exposed continental crust could have also been created, but this contingency is more highly debated. Another possibility for exposed land comes from the craters formed by large impacts, the rims of which could have risen above the surface of the ocean. The lack of dry land would have had significant implications for prebiotic chemistry as dry land is needed for many origins-of-life scenarios, which we will discuss further in later chapters.

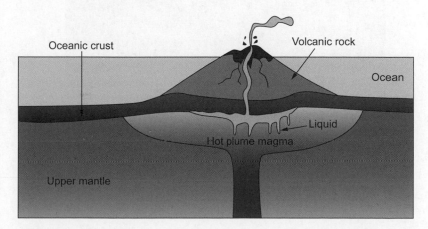

Figure 2.16 Depiction of how hotspot volcanic islands form from mantle plumes beneath the crust.

Hot springs and the possibility of wet/dry cycling

While there is no general consensus as to whether or how much dry land was available during the age of prebiotic chemistry in the Hadean, most agree volcanic islands are more likely than continental crust. The terrain surrounding the volcano would have also been geologically active, resembling *hot spring systems* we are familiar with today, which are home to a diverse array of extremophile microorganisms with different hot springs offering their own unique geochemistries, see Figure 2.17. Unlike the ocean, these smaller bodies of water are more likely to afford the accumulation of concentrated organic material produced from atmospheric processes into a 'prebiotic soup'. Small aqueous reservoirs also have the ability to undergo wet/dry cycling, which could have been useful for driving **condensation reactions** necessary for polymerization of relevant molecules like amino acids and

Figure 2.17 Photograph of a modern-day hot spring system in New Zealand.
Dry land also presents the possibility for freshwater ponds, lakes, and streams derived from rainfall.

nucleotides discussed in Chapter 7. Exposure to UV light is also possible, if not certain when considering longer timescales. Subjection to more intense tides for low elevation volcanic island systems is also something to consider.

Whether the origin of life started in the open ocean or on land is still a subject of debate, so both scenarios are likely worth investigating.

2.7 Summary

- The Solar System formed between 4.5 and 4.6 billion years ago, and Earth is thought to be ~4.56 billion years old. Little of Earth's earliest rock record has survived to the present day.

- Organic molecules could have been synthesized in space and delivered exogenously or synthesized endogenously on Earth.

- Following the Moon-forming impact, Earth acquired its secondary atmosphere likely composed primarily of H_2O, N_2, and CO_2 based on volcanic outgassing.

- Atmospheres containing gaseous molecules with more reduced oxidation states tend to be better for forming organic molecules.

- The Miller–Urey experiment demonstrated amino acid synthesis from the action of electric discharges (simulating lightning) on reduced gas mixtures and water vapour.

- Other likely energy sources include UV light, solar energetic particles, and kinetic energy delivered through cometary and meteorite impacts.

2.8 Exercises

1. Saturn's Moon Titan has lakes of liquid methane (CH_4) on its surface. What are the chemical and physical differences between liquid methane and liquid water? Do you think liquid methane could be used instead of liquid water as the solvent for life?

2. Whether life started on dry land or in the ocean is still debated. What are the pros and cons for an origin of life at a submarine hydrothermal vent versus a hot spring system on dry land?

3. The James Webb Space Telescope will be able to spectroscopically image the atmospheres of some exoplanets. What do you think the atmospheric composition of a typical terrestrial exoplanet might be?

4. The chemistry of carbonaceous meteorites suggests that the abiotic production of some important biomolecules like amino acids can occur in space on their respective parent bodies. Do you think it is possible that primitive forms of life could have arisen on meteorite parent bodies? Why or why not?

5. What mechanism do you think is more important (if any) for the inventory of organic molecules on early Earth, endogenous production or exogenous delivery?

2.9 Suggested Reading

1. Hugh R. Rollinson. (2007). *Early Earth Systems: A Geochemical Approach*. Malden, MA, USA: Wiley.

2. John E. Chambers and Alex N. Halliday. (2014). The Origin of the Solar System, in *Encyclopedia of the Solar System* (third edn). Tilman Spohn, Doris Breuer, and Torrence V. Johnson (eds). Kidlington, Oxford, UK: Elsevier, pp. 29–54.

3. Muriel Gargaud, Bernard Barbier, Hervé Martin, and Jacques Reisse (eds). (2005). *Lectures in Astrobiology Vol I*. Heidelberg, Berlin: Springer.

4. Kevin Zahnle, Laura Schaefer, and Bruce Fegley. (2010). 'Earth's Earliest Atmospheres'. *Cold Spring Harb Perspect Biol 2*(10): a004895.

5. Sukrit Ranjan and Dimitar D. Sasselov. (2016). 'Influence of the UV Environment on the Synthesis of Prebiotic Molecules'. *Astrobiology 16*(1): 68–88.

An Overview of Biochemistry

Extant life may be a useful guide to understand the origins of life. However, like any extant system that has an evolutionary history, how it came into existence may be convoluted, and some information about previous states may be lost.

A good example is language. It is estimated that ~19% of the world's population can presently communicate using English, even though it may not be their native language. English itself as a language is a mash-up of French, German, and Gaelic, and though it is hard to place strict boundaries around languages, English as a language is perhaps only 1–1.5 thousand years old. The global utility of speaking English is quite recent, and the prevalence of English speaking far outstrips the speaking prevalence of the runner-up language (Mandarin), whose usage is much more localized. Likewise, Spanish was until recently only spoken by a small percentage of the world's population but spread rapidly during Spain's colonial period which waxed and waned fairly quickly. Clearly, any given language is just one way of representing symbols about the natural environment, and people did well enough with their pre-contact systems. Whether the acquired system is a better one is debatable, nevertheless languages appear to spread like wildfire from small populations due to historical events.

The universal language of life, i.e. biochemistry, may be similarly an arbitrary representational language, spread by similar usage-advantage phenomena. Nevertheless, now that a common biochemistry among all organisms exists, we will ask questions about its origins that almost certainly miss questions about what it competed against during its development. That said, the language of biochemistry is universal enough that we recognize it as such. It may or may not offer inferences as to the earliest processes by which life emerged. At the very least, biochemistry represents an endpoint in life's origins journey, and if only for that reason, we need to study it.

This chapter will review essential biochemical and biological principles at the microscopic level. We will present a summary of the molecules common to all life, namely deoxyribonucleic acid (DNA), ribonucleic acid (RNA), proteins, and phospholipids, and their cellular functions. All life also depends on metabolic pathways, some of which are common among all organisms, and others which are unique.

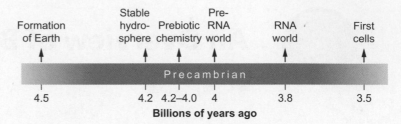

Figure 3.1 Hypothetical timeline of how life may have started on early Earth starting with its formation and ending with the appearance of the first modern cells.

3.1 A Brief History of Life on Earth

The history of life on Earth (Figure 3.1) is recorded and can be understood using the fossil record. The fossil record is the sum total of fossilized organisms and their location in the rock strata. The fossil record contains not only the fossilized remains of macroscopic plant and animal species but also of microorganisms, and the structures made by communities of microorganisms. The study of microscopic fossils is a field known as **micropalaeontology**. As a consequence of the rock cycle and plate tectonics, however, most of Earth's oldest rocks and hence the oldest fossils have been lost or metamorphosed beyond recognition. Adding to that difficulty, microscopic fossils so old have been subjected to the ravages of deep time, and so a scientific consensus is not always achieved—sometimes what appears to be fossilized remains may be just the result of abiotic geological processes. Nevertheless, there are some examples for ancient life in the fossil record which are generally accepted by the scientific community in the form of fossilized **stromatolites**.

Stromatolites and isotopic evidence for life's first appearance

Stromatolites are macroscopic sedimentary structures formed by microbial communities. As these communities grow, they deposit thin layers of (generally) calcareous (calcium based) minerals, or trap and bind calcareous sand grains, forming a stratified structure, hence the name stromatolite, from the Greek word 'stroma' meaning layer. Although rare, examples of living stromatolites still exist, for example, in Shark Bay in Western Australia. What is preserved in the fossil record is generally not the microorganisms themselves, but the mineralized structures they leave behind. Fossilized stromatolites appear throughout the fossil record, and indeed are the only fossil for almost 3 billion years of Earth history, but the oldest ones that have general scientific consensus date back to nearly 3.5 billion years ago and are also found in Western Australia, see Figures 3.2A and 3.2B.

Besides mineralized structures, scientists employ isotopic analysis of stromatolites and their contained carbon-bearing materials (carbonate minerals and carbon-bearing organic matter) to infer a biogenic origin. The two dominant isotopes of carbon are ^{12}C (~99%) and ^{13}C (~1%). Modern photosynthetic organisms uptake inorganic carbon, namely CO_2 from the environment and convert it into carbohydrates. In carrying out these enzyme-catalysed reactions, the lighter ^{12}C isotope is associated with a kinetic isotope effect that favours its incorporation over ^{13}C. Hence, these organisms become enriched in ^{12}C, and this isotopic enrichment can be measured. By applying this sort of isotopic analysis on some of the world's oldest rocks, it is possible

A)

B)

Figure 3.2 A) Photographic image of living stromatolites found in Shark Bay Western Australia. **B**) Fossilized stromatolites also found in Western Australia in the Pilbara.

to conclude that life may have started as early as 3.85 billion years ago, although this claim is still strongly debated. Fossil stromatolites 3.7 Gyr old from Greenland are more promising, but still debated, whereas the evidence for life in the 3.5-billion-year-old rocks of the Pilbara in Western Australia are compelling. Taking the appearance of liquid water as a necessary condition which can be constrained at 4.2 billion years ago, we can estimate that life emerged on Earth somewhere between 4.2 and at least 3.5 billion years ago—a 700-million-year window.

The last universal common ancestor and the progenote

Given the enormous diversity of life we see around us, it is remarkable that they all share the same biochemistry based on deoxyribonucleic acid (DNA), ribonucleic acid (RNA), and proteins. This observation suggests that all organisms are related through common ancestry. Phylogenetic data also support this idea, allowing for the construction of a **universal tree of life**. At the base of this tree is the organism from which all life descended known as the last universal common ancestor (LUCA). Before LUCA at the 'roots' of the tree of life is the prebiotic chemistry from which LUCA or any of its possible predecessors, sometimes referred to as **progenotes**, emerged. LUCA was probably not the first primitive life form to arise from the prebiotic geochemical landscape, but it was the one from which all other extant life evolved, including humans. One of the goals of prebiotic chemistry is to demonstrate the chemical steps needed to arrive at LUCA or one of its more primitive progenotes that was capable of sustaining replication, see Figure 3.3.

Extremophiles

Extremophiles are a class of organisms that may provide insight into the nature of LUCA while showcasing the range of challenging conditions life is capable of thriving in. As their name stipulates, these organisms thrive in extreme environments, for example, locations of extreme heat, cold, salinity, pH, pressure, and radiation.

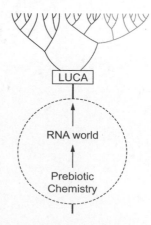

Figure 3.3 Illustration of the universal tree of life including LUCA at the base and prebiotic chemistry and the RNA world at the roots.

Extremophiles are found in all three domains of life, but most are microorganisms, with a high proportion represented by the archaea. Phylogenetic analyses have revealed that some of these microorganisms form a cluster on the base of the tree of life. One group in particular, the heat-loving hyperthermophiles, lies close to LUCA, which has led some to suggest that early progenotes may have thrived in high-temperature conditions. From an astrobiological perspective, the existence of extremophiles on Earth suggests that extant life on other less hospitable extraterrestrial locations like Mars, for example, is a distinct possibility. From a prebiotic chemistry perspective, it may suggest life could have started in a number of seemingly disparate locations.

The Archean eon and beyond

While life, including LUCA or its more primitive progenotes, may have emerged during the Hadean eon, it definitely established itself by the early Archean, the eon which lasted from 4.0 to 2.5 billion years ago. The name Archean is derived from the Greek word for 'beginning' signifying the start of the rock and fossil record (although some rocks and minerals dating from the Hadean have been found, e.g. zircon crystals discussed previously). The Archean is the age where microorganisms began to dominate the Earth, including the photosynthesizers that started producing the oxygen that eventually provided the energy for the evolution of the eukaryotic cell and the complex multicellular life forms which multiplied during the Cambrian explosion roughly 540 million years ago. The evolution of complex plant and animal life, however, is well beyond the realm of prebiotic chemistry and thus we conclude our discussion of life's history in the Archean.

3.2 The Essential Features of Prokaryotes

The fundamental unit of life is the cell. This has been evident since the 19th century. The simplest and most abundant life forms are single-celled, and microorganisms containing nuclei appear to be late arrivals based on the fossil record. Thus, the **prokaryotes** deserve special scrutiny in understanding the origins of life. Prokaryotes lack nuclei and other organelles, and the name comes from 'pro' meaning *before* and 'karyot' meaning nucleus.

Both bacteria and archaea are prokaryotes. According to the universal tree of life based on current evidence, these categories of organisms were the initial two domains of life, followed by the emergence of the more complex **eukaryotes** possessing a nucleus. Since prokaryotes, especially the archaea, are more closely related to LUCA as discussed in Chapter 1, our discussion of cellular structure will not include the eukaryotes.

Whence the origin of viruses?

The classic debate about viruses is deciding whether they are truly alive or not. As we have discussed before, it depends on one's (arbitrary) definition of life, and such definitions do little to help us understand their origins. It is true that viruses as we currently understand them depend on cellular organisms for their reproduction, which might make one consider that they must have appeared after the emergence of the first cells. However, the situation is complicated by the fact that viruses do not appear on the universal tree of life, and so pinpointing exactly when and how they emerged is difficult. Moreover, since it is widely thought that the first protocells reproduced themselves entirely without the aid of enzymes, could hypothetical primitive 'protoviruses' not have done the same? Hence, it is entirely possible that primitive viruses could have emerged early on alongside the first primitive cells. The origin of viruses is also still an unresolved scientific question, and we will not discuss it further here.

Prokaryotic cell structure

Prokaryotic cells are usually small, of the order of a micron to a few microns in diameter, and most are protected by a rigid **cell wall**. Prokaryotic cell walls are often composed of polysaccharides crosslinked by peptide chains, but the exact structure of the cell wall can vary, especially for archaea. Directly beneath the cell wall is a **cell membrane** composed of **phospholipids**. Besides phospholipids, a large fraction of the cell membrane is actually different proteins essential for metabolism, e.g. ion transporters. The cell membrane contains the contents of the cell which are known collectively as the cytoplasm. Both bacteria and archaea have similar sized ribosomes, which are smaller than those found in the eukaryotes and perhaps an indication of their more ancient origin. Although they lack a nucleus, prokaryotic DNA is concentrated in a region called the nucleoid existing usually as a single chromosome, see Figure 3.4. Prokaryotes reproduce asexually through a process called *binary fission* which results in an exact copy albeit with occasional random mutations.

The cell boundaries

Although archaea and bacteria are structurally quite similar, there are important differences worth mentioning, in particular the chemistry of the cell walls and phospholipid membranes. In bacteria, the phospholipids of the cell membrane are fatty acid esters of glycerol phosphate (Figure 3.5A). Glycerol is a linear three-carbon polyol, with one hydroxyl (OH) group on each carbon, two of which form ester linkages

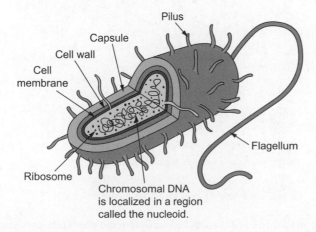

Figure 3.4 Schematic illustration of the structure and organization of a typical prokaryotic cell.

with generally unbranched **fatty acids,** and the other a phosphoester bond to a head group. These amphiphilic fatty acid phospholipids form bilayers, with the fatty acid tails pointing towards each other and the polar head groups oriented away from each other. In archaea, the ester linkage is replaced with the more stable ether linkage, and the tails are composed of isoprene-based, branched hydrocarbon structures (Figure 3.5B). In some archaea, there are also novel 'nail lipids' which are single molecule lipids with a fused tail and two heads (Figure 3.5C). All of these features add to the stability of archaeal membranes, which appear to contribute to the survivability of these organisms in extreme conditions (hence, affording them their status as ex-tremophiles). Lastly, the stereochemistry of the asymmetric carbon in the glycerol

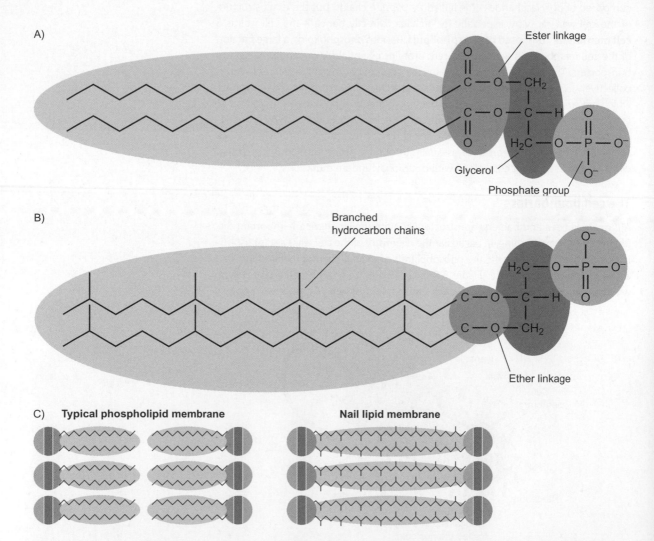

Figure 3.5 Two types of prokaryotic membrane lipids. **A)** Fatty acid acyl lipids. **B)** Archaeal isoprenoid ether lipids. **C)** A typical fatty acid lipid membrane bilayer (left) and a 'nail' lipid membrane bilayer composed of lipids with two heads connected by one tail (right).

backbone is swapped, forming the D isomer for bacteria and the L isomer for archaea. (We'll learn more about D and L stereochemistry nomenclature in Chapter 5.)

The cell walls of bacteria and archaea are also similar but vary uniquely at the molecular level. Bacterial cell walls are mostly composed of a polysaccharide known as peptidoglycan or murein. This is a linear sugar polymer appended with crosslinking short peptide chains. The polysaccharide chains are composed of repeating units of two rather unusual sugars, N-acetylglucosamine, and N-acetylmuramic acid. The short peptide chains also contain unusual amino acids, including D-isomers (the vast majority of amino acids used by cells to form proteins are L-isomers as we will learn in Chapter 6), as well as amino acids outside of the 20 that are typically found in proteins, see Figure 3.6. Gram-negative bacteria, so-called because they retain a stain developed by Hans Gram in the 19th century, visible by microscopy also possess an additional lipid bilayer outer membrane, whereas Gram-positive bacteria do not.

Cell walls are important as a layer of protection from the environment, but the added complexity that having a cell wall in addition to a membrane implies is usually taken as an indication that cell walls must have evolved after the first primitive life forms emerged. There are bacteria known as mycoplasmas that lack a cell wall entirely and only possess a cell membrane. These bacteria, however, typically make their homes inside the stable, protected environments of multicellular eukaryotic organisms, and so a cell wall is not needed. In fact, because mycoplasmas are generally parasites that rely on their hosts for many of their needs, some of them have the smallest known genomes, letting the host's metabolism do most of their work for them. Interestingly, in recent years, mycoplasmas have served as model organisms to understand what the minimal genome required for cellular life may be.

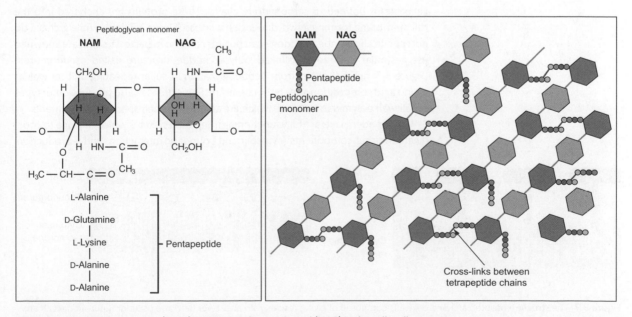

Figure 3.6 N-acetylglucosamine (NAM) and N-acetylmuramic acid (NAG) in the cell wall structure.

A minimal cell

The mycoplasma species *Mycoplasma genitalium* has one of the smallest genomes of any known modern organism. Its genome is particularly small in part because it is an obligate parasite, and its host organism maintains much of the genetic information required for its reproduction. This consideration is especially relevant in light of the heterotrophic hypothesis advanced by Oparin as it suggests organisms can be very simple provided the environment supplies a complementary amount of chemical complexity or information.

In recent years, researchers have synthesized an artificial and truncated *M. genitalium* genome in the laboratory and inserted it into a dividing bacterium of the same species creating a 'synthetic' organism to explore the minimal functional genome of a bacterium. Part of the point of these *M. genitalium* experiments was to find the limits of the non-genetically encoded requirements of life. When bacterial cells divide by binary fission, it is evident there is a great deal of information that exists ephemerally which is required for cell reproduction. For example, it is clear that to make a new cell, a genome (synthetic or natural) cannot merely be mixed with the monomers required for metabolism which would formally reproduce it. The intermediate macromolecules such as ribosomes and transfer RNA (tRNA) must also be present as we will soon discuss. On the other hand, such experiments also make it clear that happenstance co-occurrences of the appropriate macromolecular machinery and the appropriate genetic script can give rise to new organisms. The origin of viruses is thus hinted at in these experiments, but again they lay bare the problem of how a complete semantic molecular reproduction system originated in the first place.

The cell walls of archaea vary widely depending on the environment they inhabit, and some species have no cell wall at all. A large number of archaea have what is called an *S-layer* (the S stands for surface), which is composed of proteins or *glycoproteins*, i.e. proteins bonded to carbohydrate chains. These proteins are anchored into the cell membrane forming a two-dimensional lattice (Figure 3.7, left). While archaea do not specifically use the peptidoglycan polymers found in bacteria, some are known to use a similar peptide-crosslinked polysaccharide structure called pseudomurein (Figure 3.7, right). It differs from peptidoglycan in its sugar repeating units as well as amino acid composition and stereochemistry. Some archaea also produce other types of cell wall polymers/structures which are adapted to their specific environments.

While modern cells are fully functioning integrated systems which manage to coordinate their subcomponents' (cytosolic and boundary) interactions and reproduction,

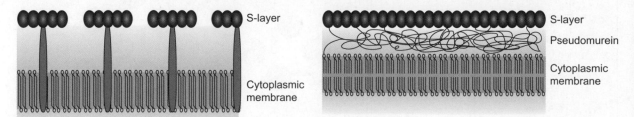

Figure 3.7 The structure of some archaea cell walls composed of glycoproteins with an S-layer (left) and pseudomurein along with an S-layer (right).

scientists attempting to understand the origins of life have developed the concept of *protocells*, which are the subject of Chapter 8.

3.3 DNA, RNA, and Proteins

All modern organisms rely on deoxyribonucleic acid (DNA), ribonucleic acid (RNA), and proteins, and this was likely also true of LUCA. These three classes of molecules are responsible for what we observe as heredity, i.e. the general tendency of an organism to possess the traits of its parent(s). These traits are carried and transferred by genes, which are the basic units of heredity. Genes are stored in DNA, the traits of which require RNA and proteins to be exhibited by the organism as well as copied for the purposes of cellular reproduction. Moreover, modern organisms react to environmental signals and changes—this reactivity is one of the hallmarks of life, and importantly, the ability to *learn* how to react to such signals also requires explanation in order to fully solve the origin of life puzzle.

Together, DNA, RNA, and proteins provide the foundation of extant biochemistry, and their specific roles and interrelations to each other are the subject of the next section.

The central dogma

In all modern organisms, genetic information flows from DNA to RNA to proteins, a process so fundamental to life it has been dubbed the **central dogma** of biology. According to modern understanding, organisms maintain DNA genomes that are used to construct **messenger RNA (mRNA)** molecules in a process called **transcription**. Organisms in turn read mRNA using ribosomes to make proteins in a process called **translation**. Depending on environmental signals, the cell is able to respond by making more or less of its various genetically encoded proteins as needed, see Figure 3.8.

All known modern organisms utilize the processes of transcription and translation. While these processes may seem complex, they involve a significant simplification of molecular information by the intermediary action of translator molecules, namely tRNA and the ribosome. In other words, the diverse molecular structures of nucleic and amino acids become generalized such that four ribonucleotide and 20 amino acid types can be treated essentially as 'movable type' by the macromolecules (RNA polymerases and the ribosome, respectively) that polymerize them. This information hierarchy—DNA to RNA to proteins—was worked out in the 1950s–60s and allowed scientists to comprehend the general mechanism by which the self-maintenance of organisms occurs. Nevertheless, this understanding does not explain the origin of such a mechanism or how such a system could evolve in the first place. By understanding the central dogma, we may begin to imagine and test experimentally the potential prebiotic chemistry that could have provided the foundation for the eventual emergence of something so sophisticated.

DNA replication

Replication of genetic material is crucial for the reproduction of all organisms. Long-term genetic information in all living organisms is stored in DNA (although there are viruses which store their heritable information in RNA, they are not capable of reproduction

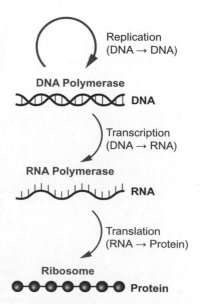

Figure 3.8 Illustration of the central dogma concept.

without a suitable host organism). DNA is a double helix composed of two single-stranded DNA polymers. Each of these polymers is composed of a linear sequence of nucleotides, which may possess any of the bases thymine, cytosine, adenine, or guanine. We will learn more about the structure and nomenclature of nucleotides in Chapter 5.

The bases are able to hydrogen bond to each other in a complementary fashion (A bonding with T, and G bonding with C). Hydrogen bonding of this type is often called Watson–Crick pairing. This hydrogen bonding motif also forms the basis of DNA copying, or replication. Replication is typically initiated by environmental signals which indicate that there are sufficient resources in the environment to enable cell division. The basic mechanism of replication is strand separation followed by copying of each strand, see Figure 3.9. During replication, the DNA duplex is unwound and pulled apart by protein enzymes, then copied one nucleobase at a time by sequential enzyme-catalysed polymerization of the complementary nucleoside triphosphates. Each separated strand serves as a **template** for the synthesis of a new complementary strand. The chemical energy stored in the triphosphates drives the polymerization. The enzymes that catalyse template-directed polymerization are known as DNA polymerases, which also have a 'proofreading' function that checks to see if the correct nucleotide has been added in the sequence before moving on. In addition to the enzymes which replicate DNA, there are a host of enzymes which continuously attempt to detect and repair DNA mutations, which constantly occur as the result of environmental and chemical mutagens, including ultraviolet (UV) light.

Transcription of DNA into mRNA

Protein biosynthesis begins with the transcription of DNA-encoded gene sequences into linear mRNA molecules, see Figure 3.10. Transcription occurs via DNA-dependent RNA polymerases, which use DNA as a template to polymerize RNA

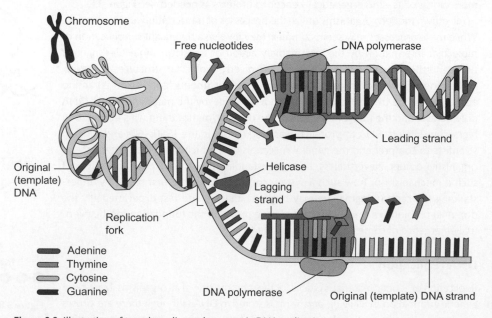

Figure 3.9 Illustration of template-directed enzymatic DNA replication.

from ribonucleoside triphosphate monomers. As in DNA replication, the chemical energy stored in nucleoside triphosphates drives polymerization forward. Unlike DNA replication, however, only a section of one of the DNA strands, i.e. the template strand, is transcribed into mRNA. The resulting mRNA is thus the complement of the DNA sequence that encodes it. There are various sequence signals in the DNA which tell the proteins needed to produce mRNA transcripts where to begin and stop copying the gene sequence, and there are various sets of which help up- or down-regulate transcription.

Translation of mRNA into proteins

The process of translation converts the information contained in mRNA sequences to corresponding peptide sequences and ultimately proteins. Besides mRNA, translation requires two other types of RNA: **ribosomal RNA (rRNA)** and **transfer RNA (tRNA)**.

Having transcribed the gene sequence into an mRNA molecule, the mRNA diffuses to the ribosome, which as discussed in Chapter 1 is mostly made of (ribosomal) RNA but also contains complexed proteins. The ribosome is what actually moves along the mRNA chain and catalyses the synthesis of polypeptides based on the mRNA sequence being translated. The ribosome brings the mRNA molecules together with the correct complement of tRNA molecules. Each specific amino acid is covalently bonded to its cognate tRNA, which transfers its amino acid to the growing peptide chain during translation. tRNA covalently bonded to its amino acid is called *aminoacyl tRNA* or 'charged' tRNA. Importantly, the information in the mRNA is read out in triplets, that is, a set of three bases known as a *codon* specifies which amino acid should be inserted in the growing peptide chain produced by the ribosomal translation complex. Each tRNA has an *anticodon* that base-pairs with the complementary codon in the mRNA, thus correctly adding the next amino acid in the sequence.

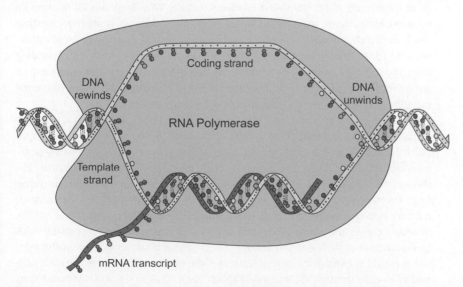

Figure 3.10 Cartoon illustration of the enzymatic transcription of DNA into mRNA.

Second letter

		U		C		A		G		
1st letter	**U**	UUU UUC	Phe	UCU UCC	Ser	UAU UAC	Tyr	UGU UGC	Cys	**U** **C**
		UUA UUG	Leu	UCA UCG		UAA UAG	Stop Stop	UGA UGG	Stop Trp	**A** **G**
	C	CUU CUC	Leu	CCU CCC	Pro	CAU CAC	His	CGU CGC	Arg	**U** **C**
		CUA CUG		CCA CCG		CAA CAG	Gln	CGA CGG		**A** **G**
	A	AUU AUC	Ile	ACU ACC	Thr	AAU AAC	Asn	AGU AGC	Ser	**U** **C**
		AUA AUG	Met	ACA ACG		AAA AAG	Lys	AGA AGG	Arg	**A** **G**
	G	GUU GUC	Val	GCU GCC	Ala	GAU GAC	Asp	GGU GGC	Gly	**U** **C**
		GUA GUG		GCA GCG		GAA GAG	Glu	GGA GGG		**A** **G**

(3rd letter shown in rightmost column)

Figure 3.11 The universal genetic code. The genetic code specifically assigns amino acids to mRNA codons.

With very few variations, all organisms use the same genetic code to perform translation, see Figure 3.11.

In bacterial translation, all peptides begin with the same 'start' codon, which encodes *N*-formylmethionine. It has been speculated that *N*-formylation prevents cyclization of the nascent peptide chain. It should be noted that since the code is structured around codons, this would allow for the encoding of 64 amino acids, yet only 20 are commonly encoded. This means the code is redundant—multiple codons often encode the same amino acid. This has been argued to be an error-buffering mechanism which ensures that relatively frequent DNA-level single base mutations do not adversely affect translated protein structure. Why there are 20 commonly encoded amino acids, as opposed to say 10 or more than 60, is another question which demands consideration of the minimum structural diversity of amino acids required for protein folding and function. It is generally believed that such coding originally specified less than 20 amino acids, and the modern set is an outgrowth of some simpler set of amino acids. The intertwining of such functional issues with error buffering questions is likely a hallmark of evolutionary processes.

tRNA plays such an important role for peptide synthesis, it deserves a bit more discussion. tRNA molecules are RNA oligomers 76–90 nucleotides in length which have large amounts of internal base complementarity and are heavily post-transcriptionally modified. The molecules fold into complex three-dimensional (3D) shapes which include two important motifs: the 3′-acceptor stem and the anticodon loop. The tRNA anticodon loop interacts directly with mRNA codons via complementary hydrogen bonding. The 3′-OH group of the terminal nucleotide of a tRNA molecule carries the amino acid. Each amino acid is attached to its cognate tRNA by a particular protein enzyme known collectively as tRNA aminoacyl synthetases, which are able to recognize various features of the tRNA and attach the correct amino acid as an ester through its carboxylate group. Each tRNA or set of tRNAs thus has its own cognate set of tRNA aminoacyl synthetases. The ability of each tRNA aminoacyl

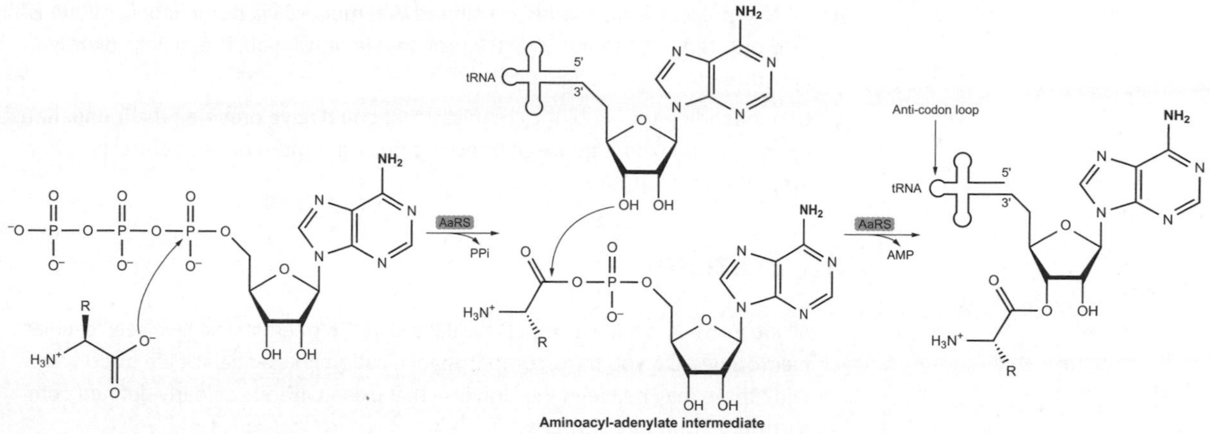

Figure 3.12 Enzymatic synthesis of charged tRNA. AaRS stands for the set of aminoacyl tRNA synthetases specific to each tRNA and its cognate amino acid.

synthetase to recognize its cognate tRNA is so important it is sometimes referred to as the **second genetic code**, see Figure 3.12.

At this point you may be asking yourself, if proteins, i.e. tRNA aminoacyl synthetases, are required for translation, then isn't that a strike against the RNA world hypothesis? If so, you are not alone. It has been hypothesized that ribozymes which could carry out the function of tRNA aminoacyl synthetases did actually exist naturally during the RNA world but were lost over time as they were replaced by more efficient aminoacyl synthetase proteins. In fact, researchers using modern laboratory techniques have discovered synthetic ribozymes that can catalyse the synthesis of charged aminoacyl tRNA, thus eliminating the need of proteins in this step altogether.

Besides rRNA, there are other naturally occurring catalytic RNA ribozymes. One well known class catalyse a process called *splicing*, in which an internal segment at a specific RNA is cut and removed, and then the two ends are re-joined. Splicing can occur en route to producing functional mRNA in both eukaryotes and prokaryotes. There are even examples of self-splicing RNA. It has also been found that the untranslated tails of several mRNA transcripts encode binding sites for the products of the metabolic pathways their gene products are involved in, which inhibit their translation. This presents yet another level of gene regulation at the level of RNA. Together, the existence of both unnaturally and naturally occurring ribozymes is generally taken as additional evidence of the RNA world hypothesis.

3.4 Summary

- The unit of life is the cell. Prokaryotes (bacteria and archaea) are the simplest and most abundant life forms on Earth, and probably the oldest.
- In all modern organisms, genetic information flows from DNA to RNA to proteins, a process called the *central dogma of biology*.

- mRNA molecules are produced using DNA templates via *transcription*. mRNA is read out using ribosomes to make proteins via *translation*. It remains unknown how these processes originated.
- Understanding the prebiotic chemistry that could have provided the foundation for the eventual emergence of transcription and translation is a central problem for prebiotic chemistry.

3.5 Exercises

1. All life as we know it is compartmentalized at the microscopic level, e.g. cellular membranes. Do you think compartmentalization is essential for life or do you think there may be life in the Universe that exists without a clearly defined compartment boundary? Explain.

2. How long do you think it took for LUCA to arise on Earth after the Moon-forming impact?

3. The RNA world likely existed before the appearance of LUCA, but do you think RNA was also the original genetic/catalytic polymer to arise spontaneously through geochemistry? Why or why not?

4. Do you think the genetic code is arbitrary, or do you think there are fundamental chemical reasons why life assigned certain codons to specific amino acids?

5. DNA and RNA are very similar in structure, yet DNA is responsible for the storage of genetic information in modern cells. Why might DNA be advantaged over RNA for this task?

3.6 Suggested Reading

1. Baker, B. J., De Anda, V., Seitz, K. W., et al. (2020). 'Diversity, Ecology and Evolution of Archaea'. *Nat Microbiol* 5: 887–900.

2. Carl W. Woese. (1987). 'Bacterial Evolution'. *Microbiol Rev 51*(2): 221–271.

3. Crick, F. (1970). 'Central Dogma of Molecular Biology'. *Nature* 227: 561–563.

4. Weiss, M., Sousa, F., Mrnjavac, N., et al. (2016). 'The Physiology and Habitat of the Last Universal Common Ancestor'. *Nat Microbiol* 1: 16116.

5. Higgs, P. and Lehman, N. (2015). 'The RNA World: Molecular Cooperation at the Origins of Life'. *Nat Rev Genet* 16: 7–17.

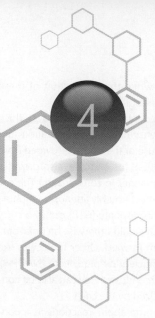

4 An Overview of Metabolism and Reaction Networks

This chapter will explore the concepts of metabolic reaction network organization. While individual organic reactions and the synthesis of specific compounds were undoubtedly important for the origins of life, these must have become organized into connected networks which helped convert environmentally supplied compounds into the components of the network, which stored and processed information about the network, catalysed the network and helped maintain its components in close spatial proximity. Tibor Gánti's 'chemoton' model for the minimal living system is a good gateway into this concept, and we will use this notion to connect prebiotic chemistry to modern biochemistry. We will briefly explore the layout of modern metabolism and explore the notion of the importance of feedback catalysis and inhibition on reaction networks for their organization. Finally, we will examine some extant prokaryotic central metabolic pathways. The hypothetical cellular life form from which all life descended, the last universal common ancestor (LUCA), and what can be inferred about its metabolism will be described.

4.1 Introduction to Metabolism

In Chapter 3, we provided an overview of the central dogma of biology. There is more to life, however, than just the central dogma. The network of reactions occurring within cells which affords the synthesis and breakdown of the materials life needs (e.g. amino acid and nucleotide building blocks), as well as the energy producing reactions needed for general cell maintenance also matter. Cells carry out an incredible number of coordinated reactions in this manner. The biosynthetic pathways, which help interconvert the material composition of the cell as well as provide energy for cell function are collectively referred to as *metabolism*.

We have not yet provided a definition of **metabolism**, so we give one here: *metabolism is the sum total of chemical reactions occurring within the cell that are essential for sustaining life*. The word metabolism derives from the Greek *metabolē*, which means 'change'. The chemical reactions of metabolism can both degrade and build up new molecules, the former process being *catabolism* and the latter *anabolism*. The break-down of **lipids** and proteins into simpler molecules like CO_2, water, and ammonia are examples of catabolism, while the production of these same macromolecules from simpler precursors is anabolism.

The chemical reactions of metabolism are organized into specific pathways, which are typically (but not always) catalysed by protein enzymes. This organization into pathways is crucial for allowing efficient capture and use of the energy provided from and required for different transformations. (Combustion, for example, also provides energy, but in a much more rapid and uncontrolled manner in comparison.) The cell spends and saves this energy in chemical 'currencies', most notably adenosine triphosphate (ATP), but the structurally related guanosine triphosphate (GTP) can also carry out similar energy-currency functions. Reactions which would otherwise be unfavourable can be coupled to ATP **hydrolysis** to drive them forward. Other molecules like nicotinamide dinucleotide (NADH), nicotinamide dinucleotide phosphate (NADPH), and flavin adenine dinucleotide ($FADH_2$) which all store redox energy can also be considered another type of chemical energy currency utilized by the cell.

While each organism has its own particular set of metabolic reactions as a consequence of its evolutionary ancestry, there are some core pathways which are common to nearly all organisms, albeit with some modifications depending on the organism in question. In this chapter, we will briefly survey some of these central prokaryotic metabolic pathways that provide precursor metabolites for other pathways, see Figure 4.1. We will first begin with a theoretical model of cellular organization.

4.2 The Chemoton Model and the Organization of Cellular Metabolism

We don't yet have a scientifically validated theory of life. We cannot predict with any certainty whether life must take the chemical and metabolic forms we find on Earth today, or if entirely different chemistries and metabolic organizations are possible. Nevertheless, in the 1970s, the theoretical biologist Tibor Gánti began publishing his work on a model for how all life, independent of its specific chemistry, must be organized at the cellular level. He referred to this model as the 'chemoton', which is short for 'chemical automaton'. In this section, we will describe this model and discuss its usefulness as a heuristic concept for understanding cellular organization.

The chemoton model of metabolism

In Chapter 1, we discussed how attempting to define life has limited utility in the sense that such definitions do little to help us understand how the transition from non-life to life takes place or what the minimum requirements for life must be. The **chemoton** model instead provides a set of minimum organizational criteria that a system must have in order to be considered as a living entity. This minimum level of organization is shown in Figure 4.2. Note that the chemoton is only an abstract system of reactions and is not defined in terms of specific chemistries.

The chemoton consists of three self-reproducing (autocatalytic) chemical cycles. First there is the cycle A → 2A, which represents the central metabolic hub of the cell. This cycle must be able not only to make all the molecules needed to reproduce the cycle, but it must also supply the compounds needed to reproduce the other two chemical systems utilizing molecules from the environment as 'nutrients'. The second

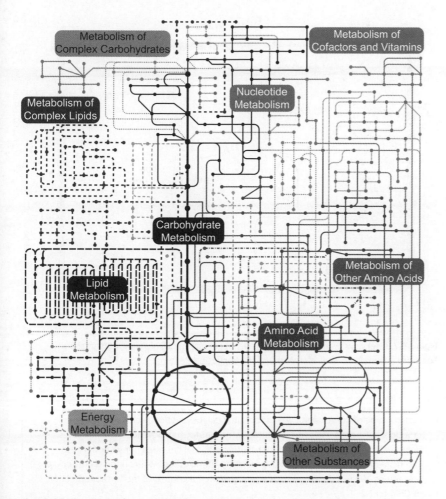

Figure 4.1 Overview of core metabolic pathways.

subsystem is the membrane cycle $T_m \rightarrow 2T_m$, which provides the boundary for the cell that separates it from the environment. The third one is the cycle $pV_n \rightarrow 2pV_n$, which represents the template-directed replication of polymeric macromolecules capable of storing information and carrying out useful functions. This cycle is also coupled to the membrane cycle such that all three work in concert as a unified metabolic system, see Figures 4.2 and 4.3.

What is autocatalysis?

Autocatalysis and autocatalytic reaction cycles are very important concepts in prebiotic chemistry and origins-of-life studies. The definition of autocatalysis is quite simple—it is catalysis done by a catalyst which is also a product of the reaction. Although seemingly definitionally simple, the implications of autocatalysis are far reaching. For example, autocatalysis implies a mechanism for chemical reproduction— you start off with one molecule of catalyst and you end up with two at the end of the reaction (since by definition a catalyst is not consumed). Biochemical DNA replication

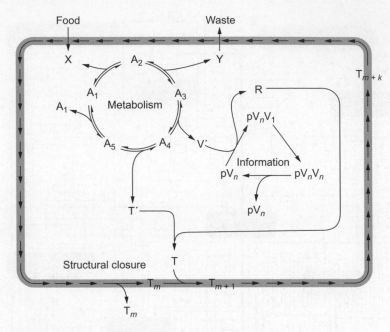

Figure 4.2 Overview of the chemoton model of cellular metabolism and organization. For more information, see reference 2 of the Suggested Reading.

is also a type of autocatalytic reaction. Autocatalytic reaction cycles also have unique kinetic properties not exhibited by non-autocatalytic reactions. Their rates become faster and faster until all the starting material has been consumed. These kinetic features provide possibilities for many 'life-like' behaviours under appropriate conditions. We will discuss autocatalysis again in the context of the formose reaction in Chapter 5.

According to Gánti, this abstract chemoton model is the minimal system which allows for what he postulates are the absolute criteria for life to be satisfied, which are the following:

First, a living system must be an individual unit, i.e. its subsystems cannot be further divided (separated) with the expectation that the unit will continue functioning as a living system. In other words, the system as a whole displays emergent properties that are more than just the sum of its parts, so separating these subsystems will result in the loss of its emergent properties and hence life-like features. Here you can see the crucial role that the membrane plays in keeping the other two subsystems working together as a single unit.

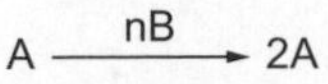

Second, a living system has to be capable of metabolism. That is to say, it must be capable of taking material and energy from the environment and

$$A \xrightarrow{nB} 2A$$

Figure 4.3 Generic scheme of simple autocatalysis, in which n number of B substrates react with A to form two molecules of A.

transforming these materials chemically into its own components, while getting rid of waste products. The chemoton model exhibits this property by all three subsystems, but especially by the central metabolic cycle A → 2A.

Third, a living system must be inherently stable. That is, despite changes in the surrounding environment, the living system's internal dynamics must remain constant. This sort of stability is related to the concept of **homeostasis**. In order to maintain a constant internal environment, the living system must be able to sense and respond to the external world.

Fourth, a living system must have a way of storing information which is some-how useful to itself, and it must be capable of decoding and using this stored information. This requirement is obviously made possible by the $pV_n → 2pV_n$ cycle.

Finally, processes in a living system must be able to be controlled and regulated. This criterion is connected to the requirement for inherent stability and metabolism, since control and regulation could not happen without these two criteria first being met.

The chemoton model is useful insofar that it provides a starting point for further experimental and theoretical development of minimal (proto)cellular systems. With this minimal model now in hand, we'll next examine some of the basic metabolic features of the simplest types of organisms—the prokaryotes.

Basic features of prokaryotic metabolism

Prokaryotes are able to live in nearly every terrestrial environment where energy, nutrients, i.e. suitable sources of carbon, nitrogen, phosphorus, etc., and liquid water are available. These microbes are complex systems in the sense that the processes they carry out are mutually interconnected—if one process fails, the whole system fails. Prokaryotes (and microbial organisms in general) must also perform this chemical balancing act under environmental regimes where the materials they produce themselves must be somewhat paradoxically both stable and reactive at the same time. In other words, like a military outfit, the materials of the cell must be able to stand at attention until 'ordered' to react when called upon.

Scientists have explored the limits to life in terms of temperature, pH, and environmental oxidation potential, and have found that prokaryotes in particular collectively have a diverse array of metabolic capabilities that allow them to inhabit extreme environments. These microorganisms carry out a wide variety of elemental assimilation (e.g. C, N, and S fixation, metal uptake, etc.) and energy transduction metabolisms. Generally speaking, there are two classes of species based on metabolism: **autotrophs** and heterotrophs. Autotrophs are able to transform inorganic carbon like CO_2 into their biomass, whereas heterotrophs obtain their carbon from already existing organic compounds. Within the autotrophs, there are chemoautotrophs whose energy is ultimately derived from chemical reactions, and photoautotrophs, whose energy is obtained from light, for example, via photosynthesis.

What was the metabolism of LUCA?

As mentioned previously, LUCA almost certainly relied on the central dogma, i.e. the system of DNA, RNA, and proteins that we are familiar with today. But other than that, is there anything more we can say about the metabolism of LUCA? In this final section, we will briefly sum up current thoughts and speculations.

All of our information about LUCA comes from comparative phylogenetic analyses of the three domains of life—the archaea, bacteria, and eukaryotes. Traditionally, organisms including microbes were classified based on their morphological (phenotypical) features, but this approach was incongruent with microorganisms as similar cellular shapes do not necessarily correlate with a common lineage.

The advent of sequencing technology in the latter part of the 20th century allowed organisms instead to be classified according to their genetic similarities. Analysing the RNA sequence of a section of the ribosome—the molecular machine which synthesizes proteins and is shared by all organisms—the microbiologist Carl Woese was able to deduce another domain of life besides bacteria and eukaryotes not previously recognized, i.e. the archaea. Since then, modern sequencing technology has allowed facile genome comparison between different organisms, enabling the identification of common sets of genes. These shared genes are thought to have also been possessed by LUCA, but even with this knowledge, there is not enough information to say for certain what LUCA was like, or what microorganism today most closely resembles it with much certainty.

Nevertheless, these phylogenetic data suggest LUCA was cellular and replicated itself using the familiar biochemistry we observe in organisms today. Some studies point to a LUCA that was anaerobic, autotrophic, made use of the acetyl-coenzyme A (CoA) pathway, was dependent on H_2 and CO_2, was thermophilic, and inhabited a hydrothermal vent environment. Even if at some point in the future, scientists could reconstruct the exact genome of LUCA with absolute certainty while pinpointing the type of environment it inhabited, the question would still remain if that was the same habitat that hosted the prebiotic chemistry from which the metabolism of LUCA emerged.

4.3 Central Metabolic Pathways in Prokaryotes

Central metabolic pathways provide the precursor metabolites for all other pathways. Some of these pathways are thought to be ancient in their origins, potentially having their roots in early Earth prebiotic chemistry. In this section, we will discuss some of the most common central metabolic pathways found in prokaryotes. Collectively, these pathways most resemble the A→2A cycle in the chemoton model.

Tricarboxylic acid cycle

The tricarboxylic acid (TCA) cycle, also known as the citric acid or Krebs cycle in honour of its discoverer, Hans Krebs, is central to metabolism, driving ATP production as well as supplying intermediates for lipid, amino acid, and nucleotide biosynthesis. In prokaryotes, the TCA cycle takes place in the cytosol, while in eukaryotes it is distributed between the cytosol and mitochondria, see Figure 4.4.

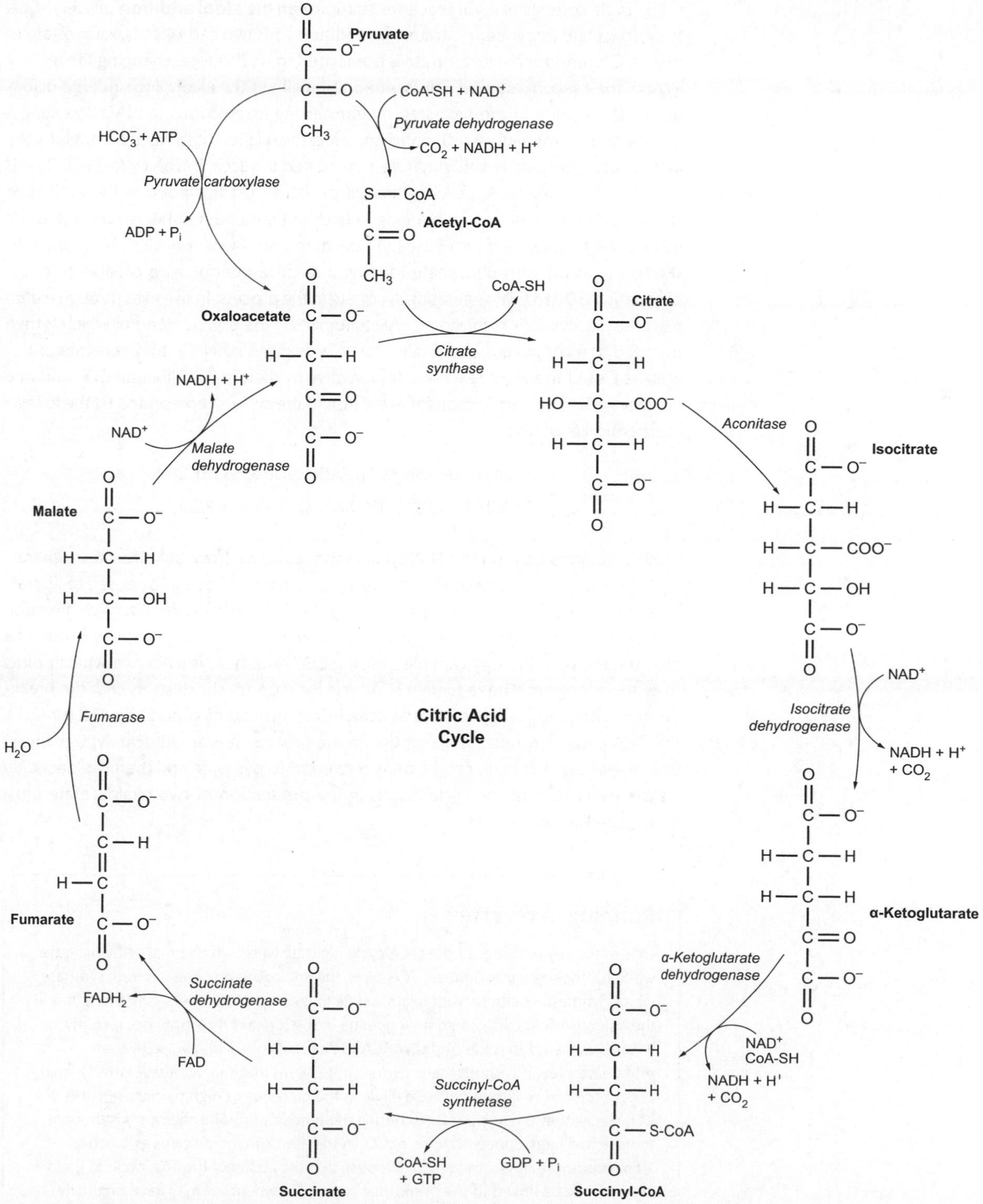

Figure 4.4 Schematic overview of the chemistry of the TCA (also known as the citric acid or Krebs) cycle.

The cycle consists of eight reactions starting with the **aldol addition** of acetyl-CoA to oxaloacetate to generate citrate. This addition adds two carbons to oxaloacetate to make a C_6 compound. Next, citrate is isomerized to isocitrate, exchanging the tertiary alcohol for a secondary one. This isomerization allows the alcohol to undergo oxidation to the ketone in the next step by transferring its electrons to NAD^+, creating a β-ketoacid intermediate, which undergoes decarboxylation to furnish α-ketoglutarate. In the fourth reaction, α-ketoglutarate is converted to succinyl-CoA by loss of CO_2 and addition of coenzyme A, CoASH, through oxidation. (At this point in the cycle, two carbon atoms have been lost, bringing us back to the same number of carbons as the initial oxaloacetate, i.e. four.) Succinyl-CoA then undergoes a nucleophilic acyl substitution reaction with phosphate forming a reactive intermediate capable of phosphorylating GDP to GTP generating succinate in the process. In the sixth through eighth reactions, succinate is oxidized by FAD to fumarate, the double bond of which is then hydrated to form malate, and finally malate is oxidized by NAD^+ to regenerate oxaloacetate. Out of the eight reactions, four involve oxidations, two liberate CO_2, and one produces GTP. The net reaction for every turn of the cycle is represented by the following chemical equation:

$$\text{Acetyl-CoA} + 3NAD^+ + FAD + GDP + P_i + 2H_2O \rightarrow$$
$$\text{CoASH} + 3NADH + FADH_2 + GTP + 3H^+ + 2CO_2$$

Some organisms can run the TCA cycle in the reductive (reverse) direction, generating the same set of intermediates, but fixing CO_2 in the process as opposed to liberating it. Microbes that are capable of running the TCA cycle in the reductive direction may be especially relevant to origins-of-life research, as this pathway is only one of a known handful in biology capable of fixing CO_2, and there is evidence which points to its ancient origin. The reductive TCA cycle (or rTCA for short) starts with the breakdown of citrate to oxaloacetate and acetyl-CoA. Instead of generating ATP (or GTP), the rTCA cycle consumes it, fixing CO_2 in the process. It is an autocatalytic pathway by virtue of the fact that acetyl-CoA is converted to pyruvate and then oxaloacetate. Hence, every turn of the cycle results in the production of two oxaloacetate molecules, see Figure 4.5.

Running in reverse

One of the key enzymes of the rTCA cycle is citrate lyase, which breaks down citrate, while for the regular (oxidative) TCA cycle, the analogous enzyme is citrate synthase, which synthesizes citrate from oxaloacetate and acetyl-CoA. It was long thought that the oxidative TCA cycle could only operate in the forward direction, but recently it was discovered that the oxidative TCA cycle can also run in reverse in some prokaryotes (even without the enzyme citrate lyase) that can express relatively large concentrations of citrate synthase while in the presence of high partial pressures of CO_2. Recall that the ancient Earth during the time of prebiotic chemistry is thought to have had large concentrations of CO_2 in the atmosphere, perhaps as much as 10 bars or more. It is intriguing to suppose that the ability of the TCA cycle to go in reverse may be linked to the primordial conditions whence it may have originally emerged.

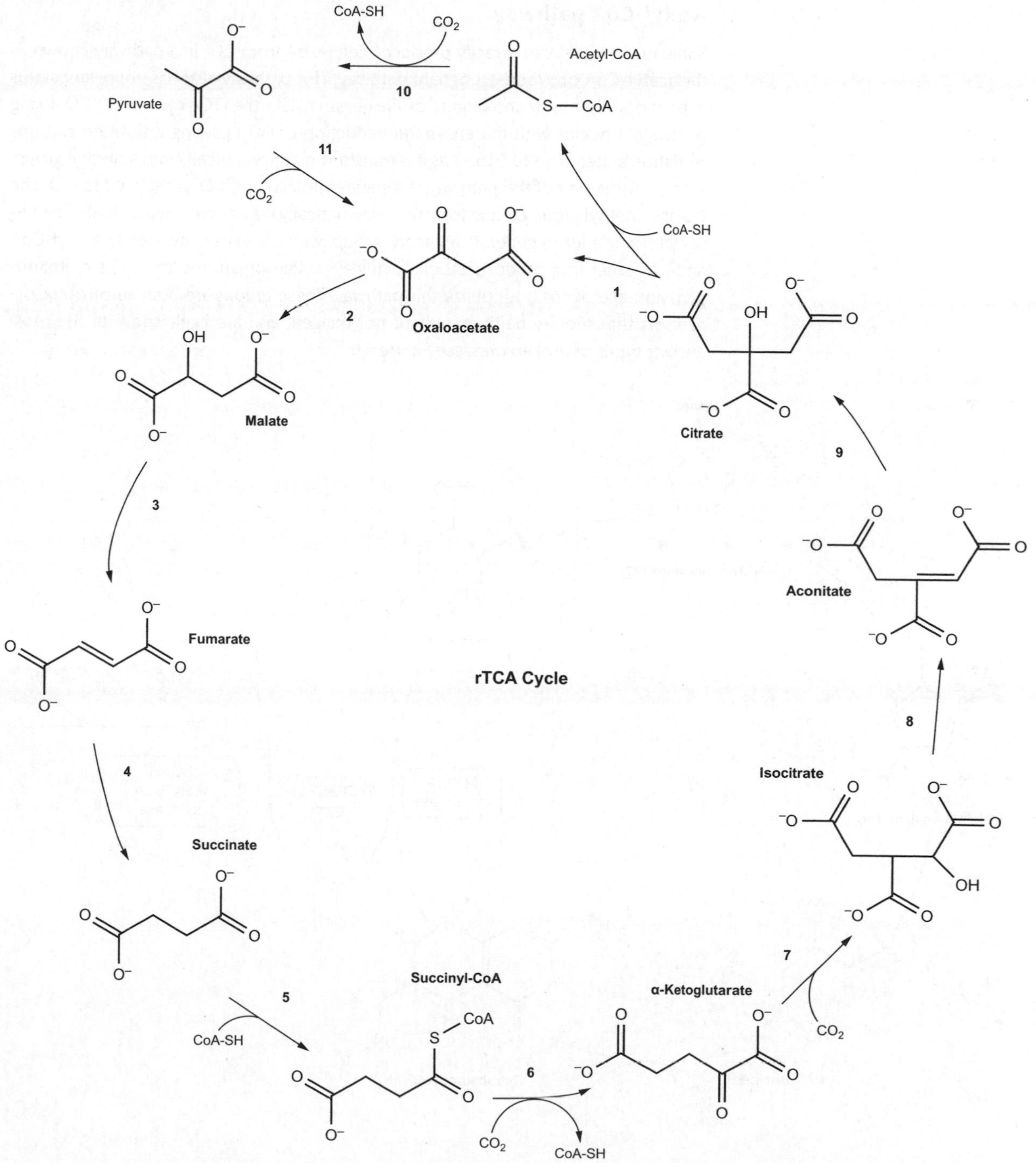

Figure 4.5 Schematic overview of the rTCA cycle. Note that one citrate is transformed into two oxaloacetates each turn of the cycle. For more information, see reference 5 of the Suggested Reading.

Acetyl-CoA pathway

Some prokaryotes can directly produce acetyl-CoA from CO_2 in a pathway known as the acetyl-CoA or Wood–Ljungdahl pathway. This pathway also has important status in prebiotic chemistry and origins-of-life research. Like the rTCA cycle, it is a CO_2 fixing pathway. It begins with the enzymatic reduction of CO_2, passing it through carbon-oxidation states of +2 to 0 to –3 as it is transformed enzymatically into a methyl group. Along another leg of the pathway, a separate molecule of CO_2 is reduced to CO. The CO and methyl group couple together enzymatically via a cobalt-dependent enzyme-cofactor complex to generate an acetyl group, which is then converted to acetyl-CoA and can enter into other metabolic pathways. Microorganisms known as methanogens and acetogens both utilize this pathway. These groups are both autotrophic organisms that require neither sunlight nor oxygen, and are likely some of the most ancient types of prokaryotes, see Figure 4.6.

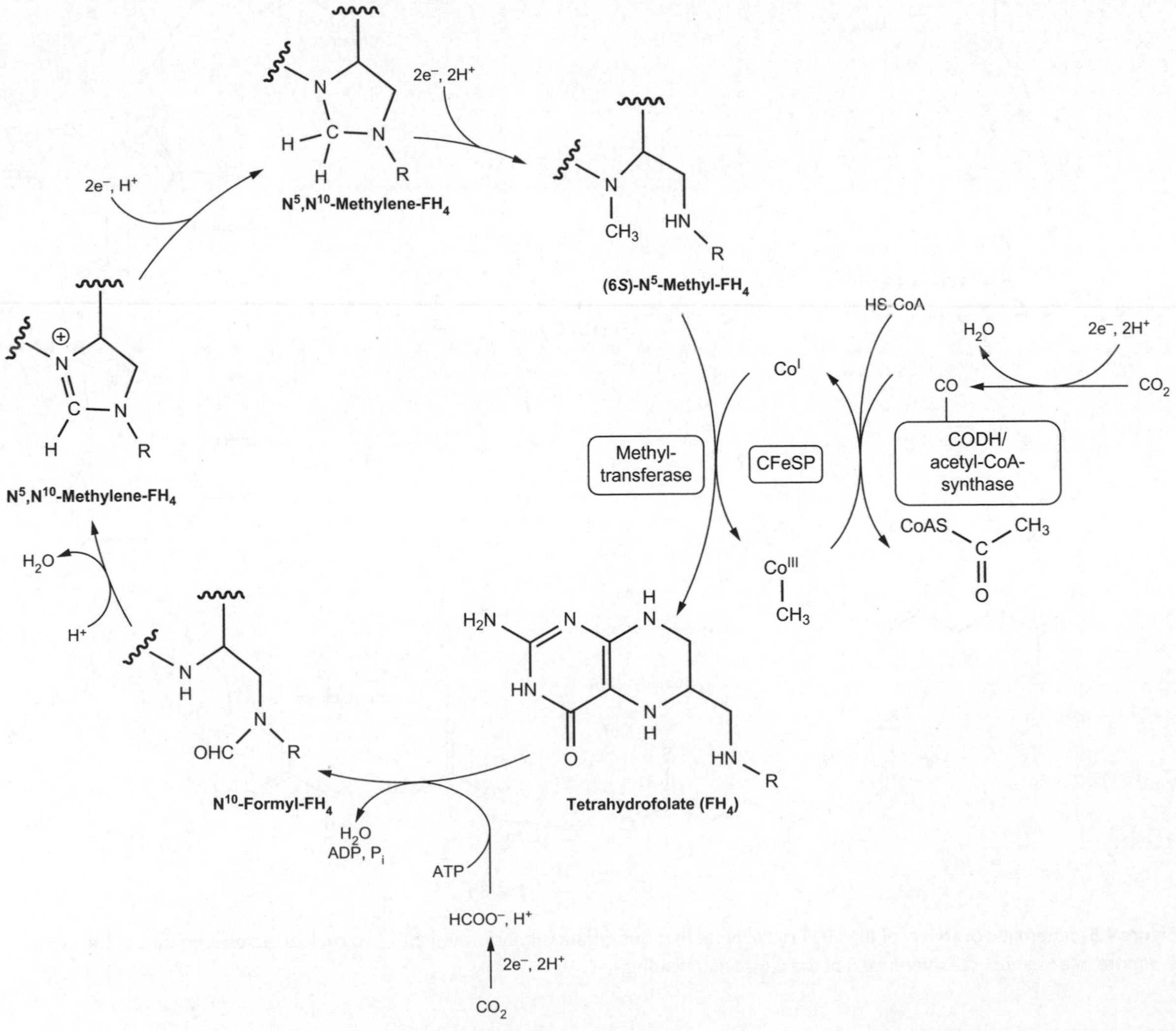

Figure 4.6 Schematic overview of the acetyl-CoA pathway.

Glycolysis and the pentose phosphate pathway (PPP)

Acetyl-CoA, which is a necessary feedstock for the TCA cycle can also be derived from the acetyl group of pyruvate together with coenzyme A. While the conversion of pyruvate to acetyl-CoA is typically discussed in biochemistry as happening under aerobic conditions, some prokaryotes carry out this transformation anaerobically. A major source of pyruvate is carbohydrate metabolism, namely through glycolysis and the pentose phosphate pathway (PPP).

Glycolysis starts with glucose, a six-carbon sugar, which is converted to two molecules of pyruvate through ten enzymatic reactions. Glycolysis also yields reducing equivalents in the form of NADH and stored energy in the form of ATP. The first and third reactions are phosphorylations which consume ATP, and this is known as the 'investment phase'. The latter part of the pathway generates ATP as well as NADH, and these reactions are known as the 'payout phase'. Overall, the pathway yields a net surplus of ATP. The reactions of the glycolysis pathway are shown in Figure 4.7A. Glycolysis doesn't only break down glucose to pyruvate, it also supplies precursor metabolites for other anabolic pathways for the synthesis of lipids, amino acids and other sugars. The overall reaction is shown by the following chemical equation:

$$\text{Glucose} + 2\text{ADP} + 2\text{P}_i + 2\text{NAD}^+ \rightarrow 2\text{pyruvate} + 2\text{ATP} + 2\text{NADH} + 2\text{H}^+$$

The PPP runs parallel to glycolysis (Figure 4.7B), sharing some of the same intermediates. As the name suggests, it is responsible for the production of pentose phosphates, in particular, ribose-5-phosphate, which is required for the biosynthesis of nucleotide derivatives which are incorporated into DNA, RNA, and various cofactors. The PPP also provides a precursor for aromatic amino acid synthesis. The PPP starts with glucose-6-phosphate (G6P), and yields fructose-6-phosphate (F6P) and glyceraldehyde-3-phosphate (G3P). The F6P generated isomerizes back to G6P, however, so the net result is just production of G3P along with NADPH and CO_2, as detailed in the following equation:

$$\text{Glucose-6-phosphate} + 6\text{NADP}^+ \rightarrow 3CO_2 + \text{glyceraldehyde-3-phosphate}$$
$$+6\text{NADPH} + 6\text{H}^+$$

G3P is important for generating ATP as it undergoes oxidative phosphorylation to form 1,3-bisphosphoglycerate, which can then phosphorylate ADP. Glyceraldehyde-3-phosphate is an intermediate in both glycolysis and gluconeogenesis, which will be discussed in the next section.

In fact, while many higher organisms use glucose as an energy storage molecule, many microbes use simple polyesters known as polyhydroxyalkanoates (PHAs). These PHAs form as water-insoluble inclusions in the cytoplasm and have properties similar to some familiar synthetic polymers produced from petrochemicals. PHAs are typically produced under limiting environmental conditions, i.e. when there is an excess carbon source available in the environment but not enough of other nutrients.

4.4 Other Important Biosynthetic Pathways

Besides the central metabolic pathways, other important pathways include those needed for the synthesis of primary metabolites, such as sugars, nucleotides, amino acids, and phospholipids. These pathways most resemble the $pV_n \rightarrow 2pV_n$ and $T_m \rightarrow 2T_m$

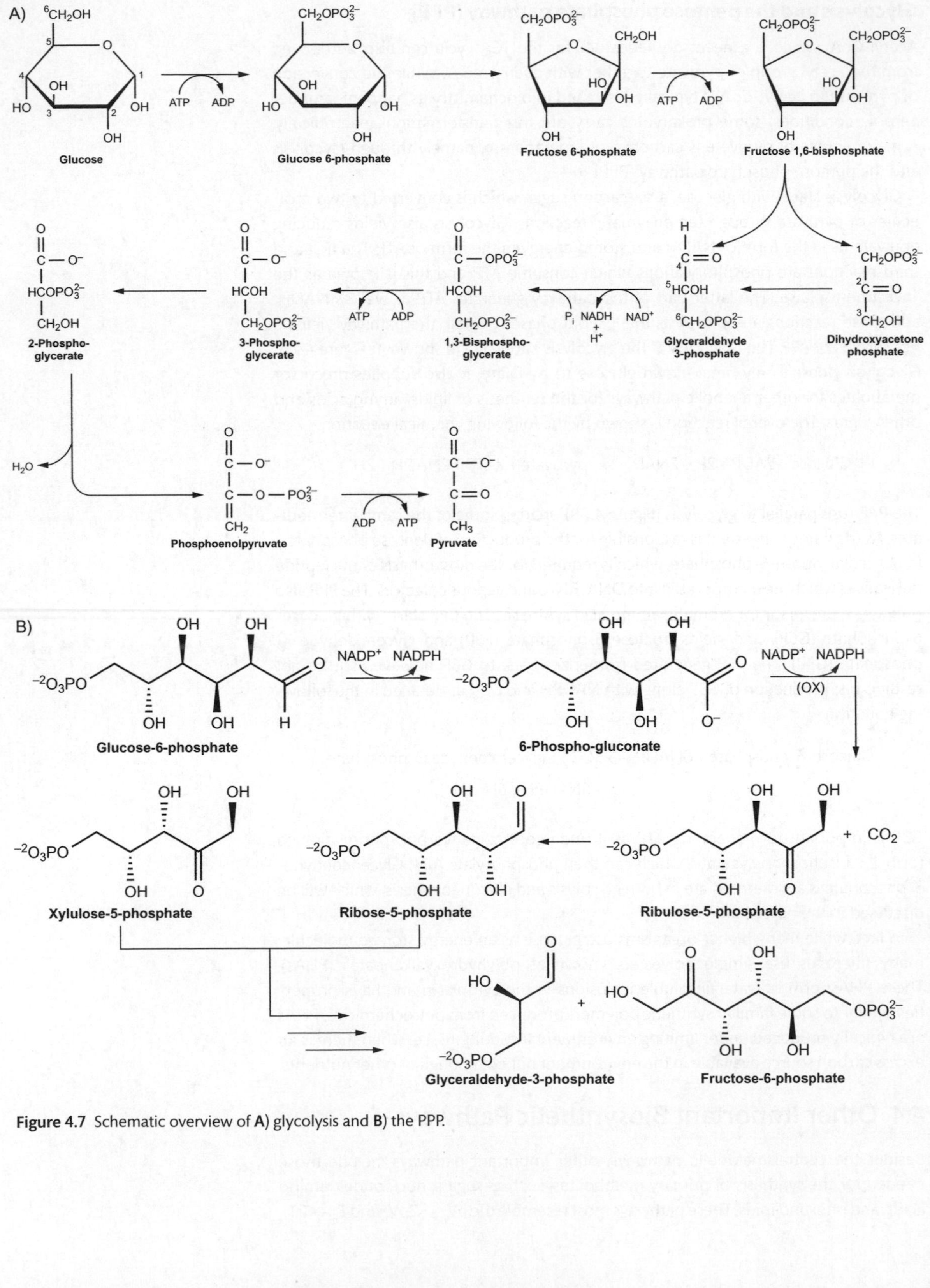

Figure 4.7 Schematic overview of **A)** glycolysis and **B)** the PPP.

cycles in the chemoton model. We briefly discuss the basic features of these in the following text.

Gluconeogenesis

While glycolysis breaks down sugars, gluconeogenesis builds them back up. Gluconeogenesis produces glucose while proceeding through many of the same intermediates and reactions as glycolysis. Gluconeogenesis is an anabolic pathway that starts from pyruvate, which seems fitting as pyruvate is the endpoint of glycolysis. Gluconeogenesis, however, is not simply the reverse of all the steps of glycolysis as some steps in glycolysis are irreversible—'detours' must be made for these steps in order for the reactions to proceed spontaneously. While glycolysis is net ATP producing, gluconeogenesis is net ATP consuming, see Figure 4.8.

Amino acid biosynthesis

Twenty species of amino acids are the building blocks of nearly all modern proteins. Ten of these are biosynthesized starting from oxaloacetate and α-ketoglutarate, which are intermediates of the TCA cycle. The other ten are synthesized from pyruvate, 3-phosphoglycerate, phosphoenolpyruvate, erythrose-4-phosphate, and phosphoribosyl diphosphate, which are intermediates of glycolysis and the PPP. Together, these three central metabolic pathways provide all the metabolites needed for amino acid biosynthesis.

The biosynthesis of glutamate and glutamine is especially crucial, as their amino groups are the source of all other N-containing compounds produced by the cell. Glutamate and glutamine are both derived from α-ketoglutarate, which undergoes reductive amination to generate a transferable amino group. The synthesis of all other amino acids proceeds by **transamination** where glutamate is the amino donor, and the substrate is an α-ketoacid, and glutamate is transformed back to α-ketoglutarate in this process. Transamination of oxaloacetate itself affords the amino acid aspartate, and transamination of pyruvate yields alanine. The other amino acids require more steps, but all involve a transamination to install the amino group, see Figure 4.9.

Nucleotide biosynthesis

RNA and DNA nucleotides are constructed from common biosynthetic pathways. The **purine** nucleotides are synthesized through one pathway, while the **pyrimidine** nucleotides are constructed through another. Both pathways make use of ribose-5-phosphate supplied by the PPP for the sugar component and a number of different amino acids to construct the bases, see Figure 4.10A and 4.10B. (For a discussion of nucleotide nomenclature and structure, see Chapter 5.)

In pyrimidine biosynthesis, aspartate, glutamine, and bicarbonate are used to construct orotate, a pyrimidine resembling uracil except substituted at the 6-carbon with a carboxylate group, which takes place over a few steps. Orotate is then reacted with 5-phosphoribosyl pyrophosphate, which itself is derived from ribose-5-phosphate, forming a **glycosidic bond** with the ribose furanosyl ring. This intermediate is further transformed over a few more steps to uridine monophosphate, uridine triphosphate and finally cytidine triphosphate. The deoxynucleotides are then generally

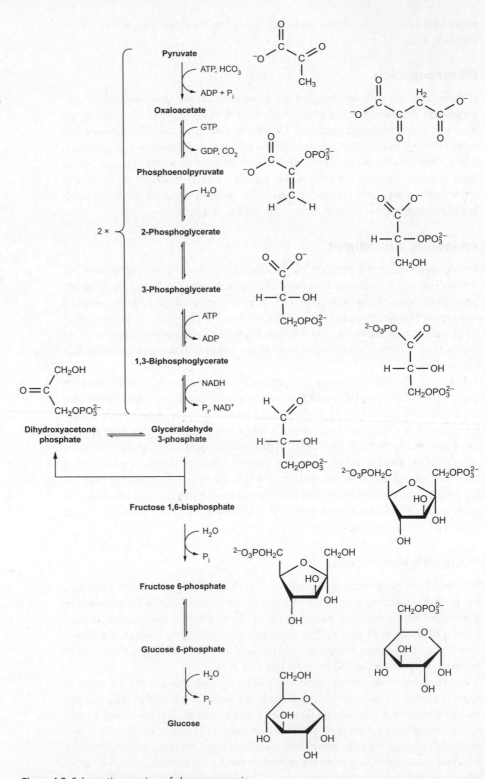

Figure 4.8 Schematic overview of gluconeogenesis.

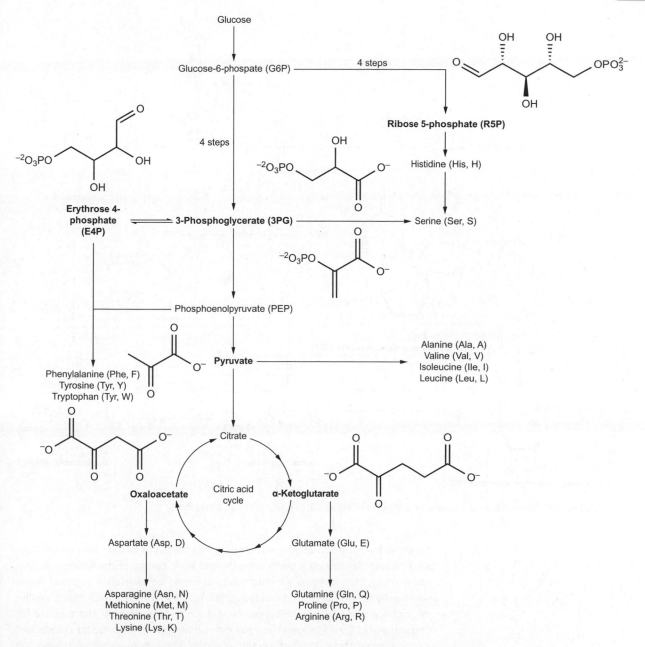

Figure 4.9 Schematic overview of amino acid biosynthesis in terms of the substrates from which the amino acid is derived.

synthesized by reduction of the 2'-carbon, and in the case of thymidine, a methyl group is installed on the uracil ring.

Purine biosynthesis also employs 5-phosphoribosyl pyrophosphate, but unlike the pyrimidine pathway, nucleobase synthesis builds out from the ribose ring as opposed to coupling a premade purine ring at a later intermediate stage. The amino acids glycine, aspartate, and glutamine, together with bicarbonate and 10-formyl-tetrahydrofolic acid (formyl-THF) are used in purine construction. Formyl-THF is a coenzyme that

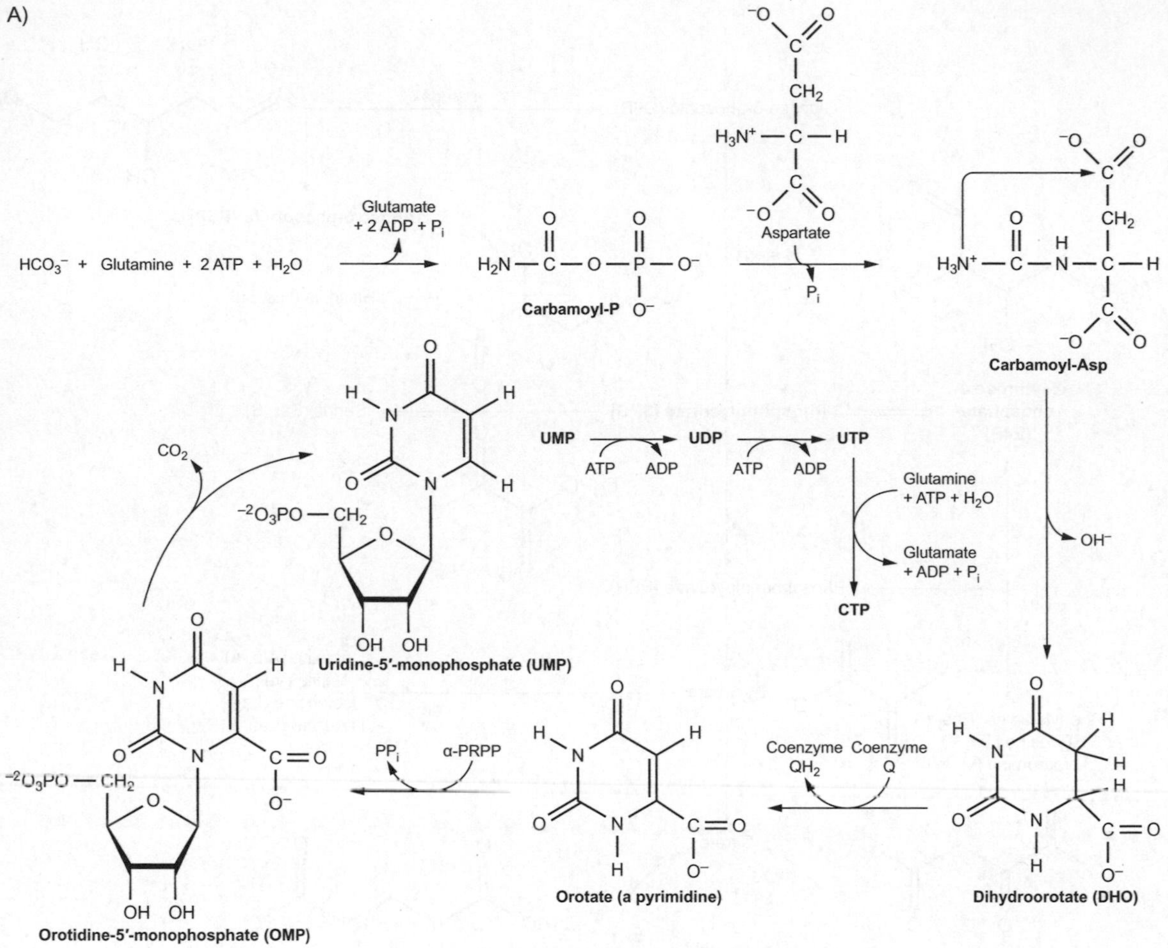

A)

Figure 4.10 Schematic overview of *de novo* **A)** pyrimidine and **B)** purine nucleotide biosynthesis.

donates a formyl group (HCO-) twice throughout the pathway. Over the first six steps, a 5-aminoimidazole ring is built up on the anomeric carbon of the ribose unit. This ring-forming step involves an intramolecular imine condensation reaction made possible by the formyl group provided by formyl-THF. Five more steps afford inosine monophosphate, which itself is a purine and serves as a common intermediate for the synthesis of both adenosine and guanosine monophosphates that take place over a couple more steps. Another intramolecular condensation reaction furnishes the fused pyrimidine ring, also made possible by the formyl group donated by formyl-THF. Just like the pyrimidine pathway, the deoxynucleotides then are made by reduction of the 2'-carbon on ribose.

Both pyrimidine and purine metabolism also employ *salvage pathways*. Rather than constructing the nucleobases *de novo* ('from scratch'), the purine and pyrimidine bases captured from degraded mRNA can be coupled directly to 5-phosphoribosyl pyrophosphate using specialized enzymes known as phosphoribosyl transferases for this exact purpose. See Figure 4.11.

B)

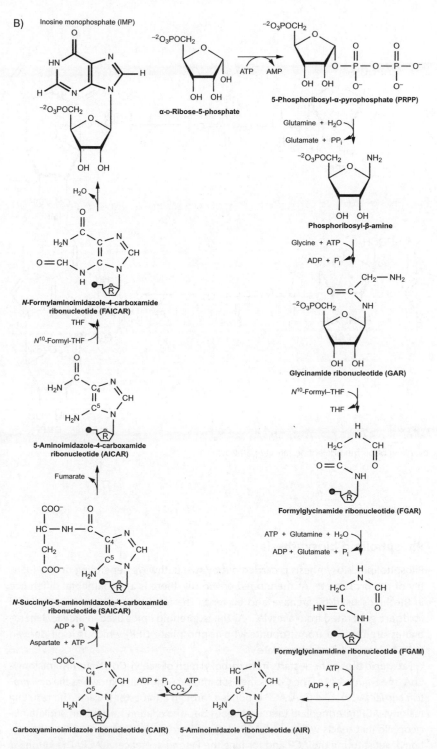

Figure 4.10 Continued

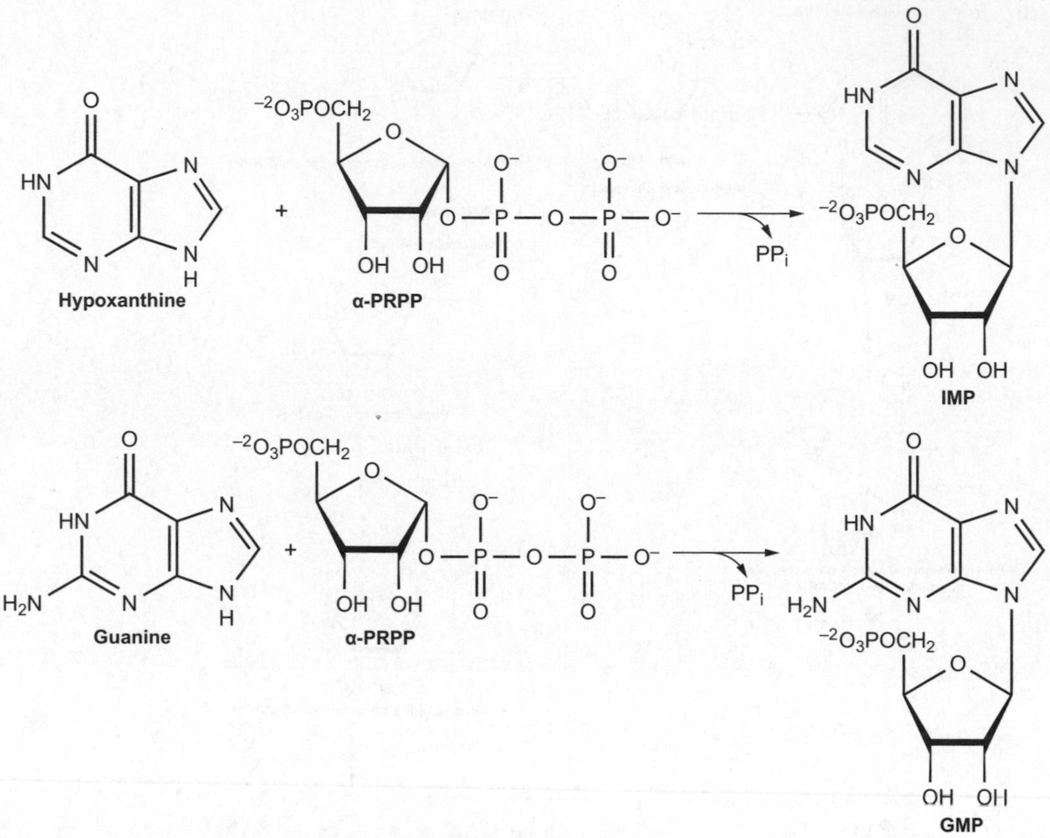

Figure 4.11 Schematic overview of purine nucleotide salvage pathways.

Phospholipid biosynthesis

Phospholipid biosynthesis is carried out by two pathways depending on the identity of the prokaryote. As mentioned previously, there is a fundamental difference in the lipids used by archaea and bacteria. The fatty acid tails used in bacterial lipids are generated from acetyl-CoA. The isoprenoid lipids used in archaeal membranes are derived from isopentenyl pyrophosphate (IPP) which is itself derived from 1-deoxyxylulose phosphate.

Fatty acid biosynthesis starts from carboxylation of acetyl-CoA to furnish malonyl-CoA, see Figure 4.12A. The CoA is then substituted with what's known as *the acyl protein carrier* (ACP). Like CoA, ACP is bonded through a thioester linkage. The resulting malonyl-ACP intermediate then undergoes decarboxylation forming an enolate nucleophile that reacts with acetyl-ACP (derived from acetyl-CoA and ACP) at the carbonyl, substituting the ACP and forming the ketone acetoacetyl-ACP. The ketone is reduced over a couple of steps to yield the saturated butyryl-ACP. This series of steps is then repeated multiple times until the correct chain length has been achieved.

For each elongation, malonyl-ACP serves as the nucleophile which attacks the carbonyl of the thioester of the growing hydrocarbon chain, displacing the ACP. If an unsaturated fatty acid is needed, then a special dehydrase enzyme comes and installs the double bond in the appropriate position at an intermediate step in the elongation process.

Once fatty acid biosynthesis has been completed, fatty acids undergo esterification with glycerol-3-phosphate making use of the ACP as a **leaving group**. Depending on the identity of the phospholipid, e.g. phosphatidylcholine, etc., the phosphate is then further functionalized.

In the case of the archaeal lipids, phospholipids are also constructed from a glycerol backbone, but with the phosphate in the 1 instead of 3 position. In fact, This swapping of the position of the phosphate makes the **asymmetric centre** of the central carbon of opposite handedness in comparison to the glycerol backbone of bacterial fatty acid phospholipids. The remaining two alcohols of the glycerol backbone are then functionalized with an isoprenoid tail through an ether linkage, specifically, the compound *geranylgeranyl-pyrophosphate* derived from *isopentenyl pyrophosphate*. After that, the phosphate becomes functionalized with the target headgroup depending on the specific biosynthesis pathway, followed by reduction of the double bonds, see Figure 4.12B.

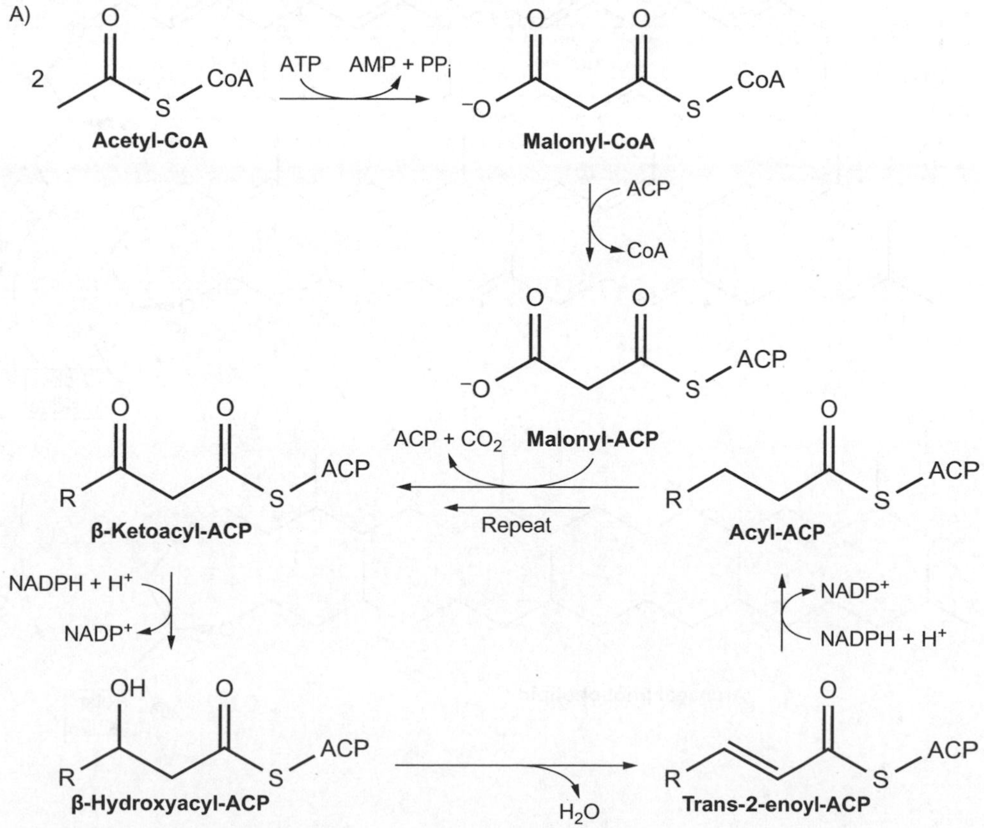

Figure 4.12 Overview of membrane lipid biosynthesis for both **A)** bacterial and **B)** archaeal phospholipids.

B)

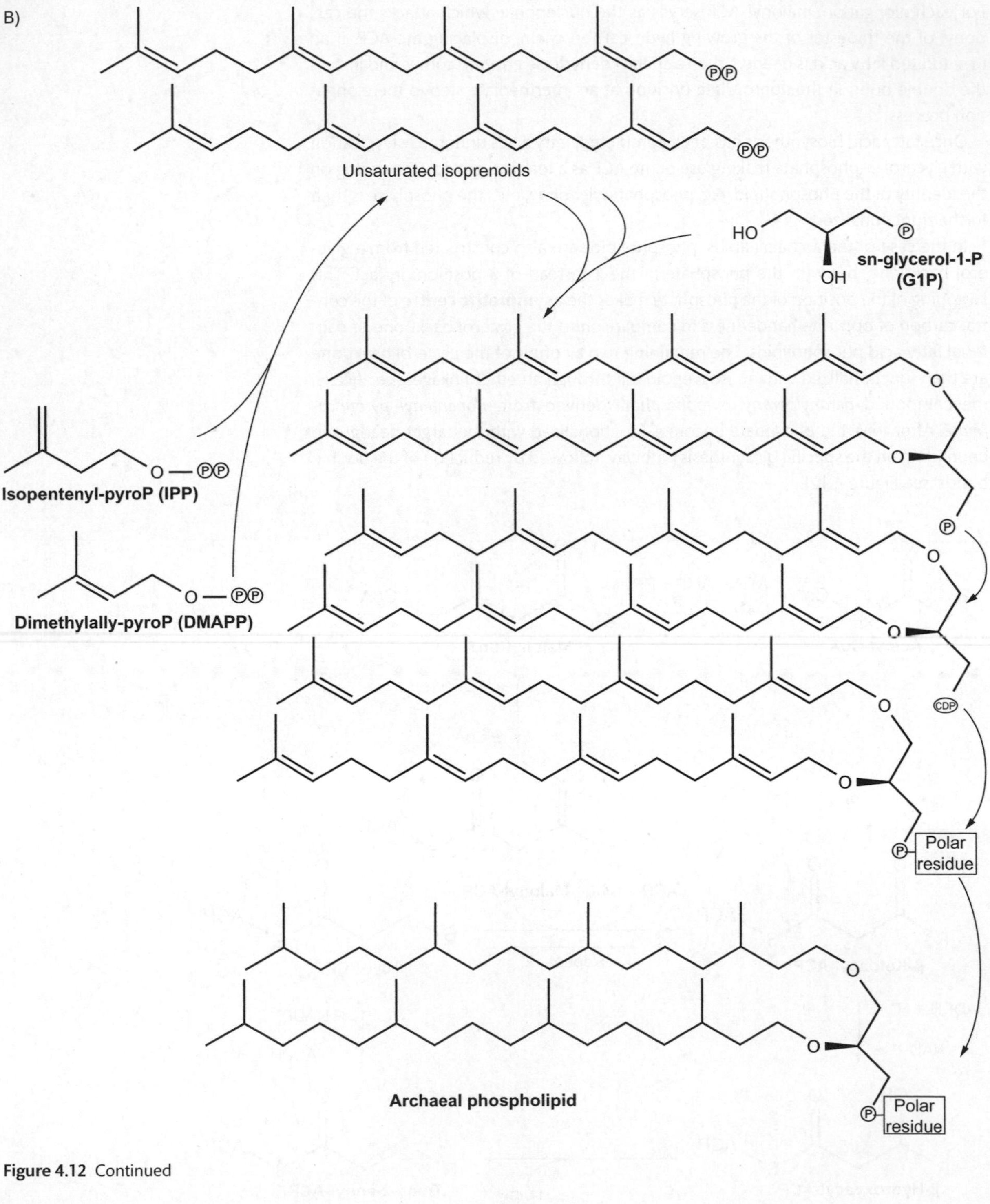
Unsaturated isoprenoids
HO
sn-glycerol-1-P
(G1P)
OH
Isopentenyl-pyroP (IPP)
Dimethylally-pyroP (DMAPP)
CDP
Polar residue
Archaeal phospholipid
Polar residue

Figure 4.12 Continued

Geranylgeranyl-pyrophosphate is a common precursor for the synthesis of compounds known generally as terpenes or terpenoids, which are generally considered **secondary metabolites**. Secondary metabolites are compounds produced by an organism which are not essential for growth, development, and reproduction, whereas **primary metabolites**, e.g. DNA, RNA, proteins, and lipids, are. For this reason, secondary metabolites are not usually considered as relevant to prebiotic chemistry, but it is noteworthy that archaea use what is often regarded as a precursor for secondary metabolites for their primary metabolism.

4.5 Summary

- Metabolism is the sum of all chemical reactions occurring within the cell that are essential for sustaining life.

- The chemoton model of the cell provides a set of organizational criteria that a chemical system must have in order to be considered as a living entity: a container, a metabolism, and an information storage capacity.

- Central metabolic pathways provide precursor metabolites for other pathways. Some of these pathways like the (r)TCA cycle and the acetyl-CoA pathway are thought to be ancient, potentially having roots in prebiotic chemistry.

- Twenty amino acids are the building blocks of all modern proteins. Ten of these are biosynthesized starting from oxaloacetate and α-ketoglutarate, which are intermediates of the TCA cycle. The other ten are synthesized from intermediates of glycolysis and the PPP (pentose phosphate pathway).

- RNA and DNA nucleotides are constructed from common biosynthetic pathways. Both pathways use ribose-5-phosphate supplied by the PPP for the sugar component and various amino acids to construct the nucleobases.

4.6 Exercises

1. Do you think it is possible the central metabolic pathways evolved independently or prior to the advent of a genetic apparatus? Why or why not?

2. Do you think the rTCA cycle can proceed chemically without the aid of enzymes, or do you think the cycle requires fundamental changes in chemistry to proceed nonenzymatically?

3. Explain in your own words what autocatalysis is and why it is a fundamental feature of the chemoton model.

4. Do you think *de novo* nucleotide biosynthesis or the salvage pathways evolved first? Explain your reasoning.

5. Biology tends to use a set of similar chemical transformations over and over again, e.g. transamination in the synthesis of amino acids. Why do you think this might be the case?

4.7 Suggested Reading

1. Eric Smith and Harold J. Morowitz. (2006). *The Origin and Nature of Life on Earth: The Emergence of the Fourth Geosphere*, 1st edn. United Kingdom: Cambridge University Press.

2. Tibor Gànti. (2003). *Chemoton Theory: Theory of Living Systems*. New York, NY: Springer.

3. Kamila B. Muchowska, Sreejith J. Varma, and Joseph Moran. (2020). 'Nonenzymatic Metabolic Reactions and Life's Origins'. *Chem Rev 120*(15): 7708–7744.

4. Eric Smith and Harold J. Morowitz. (2004). 'Universality in Intermediary Metabolism'. *Proc Natl Acad Sci USA 101*(36): 13168–13173.

5. Leslie E Orgel. (2008). 'The Implausibility of Metabolic Cycles on the Prebiotic Earth'. *PLoS Biol 6*(1): e18.

5 Sugars, Nucleobases, and RNA: Prebiotic Ribonucleotide Synthesis

This chapter marks the beginning of a deeper dive into organic chemistry. The prebiotic and organic chemistry of sugars, nucleobases, and nucleotides is discussed. We begin by explaining the nomenclature of sugar chemistry. Important stereochemical concepts are discussed, including the notion of enantiomers, diastereomers, and anomers. Possible abiotic syntheses of sugars via the Kiliani–Fischer and formose reactions are detailed. The basic chemical structures of the pyrimidine and purine nucleobases are described as well as their potential prebiotic syntheses. Examples of major strategies for prebiotic nucleoside/nucleotide assembly are given.

5.1 Sugar Nomenclature, Structure, and Stereochemistry

Stereochemistry is the branch of chemistry that deals with the spatial arrangement of atoms and functional groups in molecules, and is one of the most challenging concepts in chemistry, as it requires visualizing molecules in three-dimensional (3D) space while also memorizing a complex system of nomenclature and classifications. Sugars can have complicated stereochemistry. For these reasons, it is worth understanding the basic concepts of stereochemistry. We'll start from the beginning and work our way up to the stereochemistry of sugars.

The geometry of the carbon atom

Carbon is special among the elements in being capable of stably bonding with one, two, three or four other atoms. In the context of prebiotic chemistry, these atoms are typically hydrogen, oxygen, nitrogen, sulfur, and especially other carbons. Specifically for sugars, the atoms involved are always C, H, and O.

Any sp^3 (tetrahedral) carbon has the potential to be chiral, that is, its mirror image may not be superimposable with itself. Whenever an sp^3 carbon is bonded to four different atoms/groups, it will be chiral, that is to say its mirror image will not be

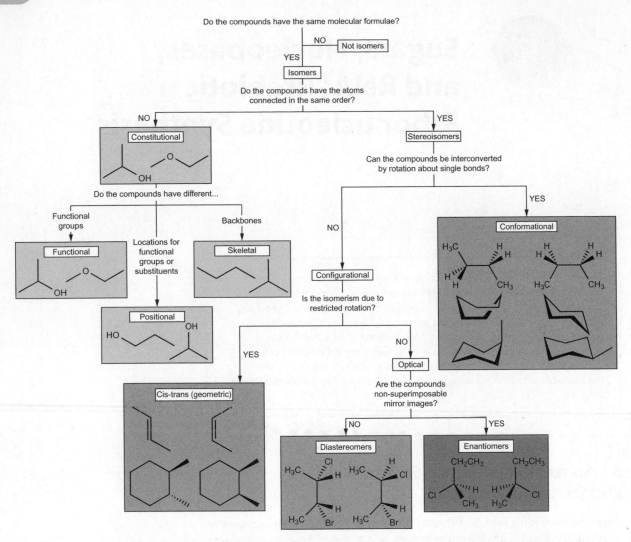

Figure 5.1 Illustration of the 'ladder' of chemical isomerism, including constitutional, conformational, geometric, diastereomeric, and enantiomeric isomers.

superimposable with itself. We give these types of carbons special names—*asymmetric* or *chiral* centres, see Figure 5.1.

Now let's focus on D-glyceraldehyde, one of the simplest sugars, as an example of a molecule with an asymmetric centre. There is only one asymmetric centre in D-glyceraldehyde, because only one carbon is both sp³ hybridized (tetrahedral) *and* bonded to four different atoms/groups. We can draw the 3D structure of D-glyceraldehyde using wedged and hashed bonds, and using this representation, we can draw its mirror image, L-glyceraldehyde (Figure 5.2). Note that D- and L-glyceraldehyde are mirror images of each other but cannot be superimposed no matter how they are rotated, so they are *different molecules*. Because they are mirror images of each other, we give them a special name—**enantiomers**. This name comes from the Greek word *enantion*, which means 'opposite'. Enantiomers are a subset of a broader class of

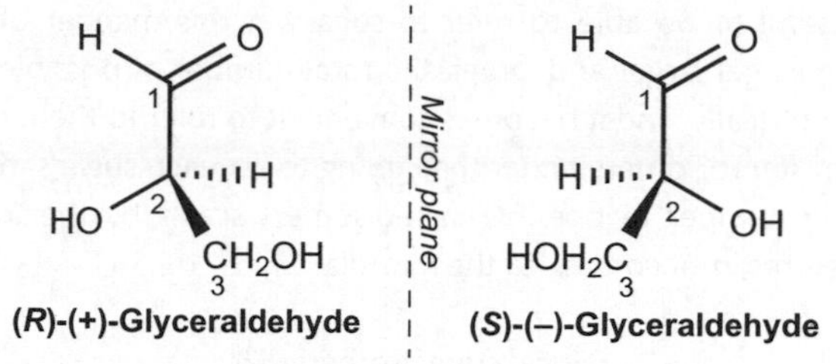

(R)-(+)-Glyceraldehyde   **(S)-(–)-Glyceraldehyde**

Figure 5.2 (R)-(+)-glyceraldehyde and (S)-(–)-glyceraldehyde showing their mirror image relationship as enantiomers and carbon numbering.

Note: The (+) and (-) signs indicate the physical property of optical activity.

isomers, known as *stereoisomers*. Stereoisomers are molecules which have the same molecular formula, the same connectivity between their constituent atoms, but with some atoms arranged differently in space.

Ketoses and aldoses

Let's now move beyond glyceraldehyde to larger sugars. Again, half of the battle when it comes to sugar chemistry is mastering the nomenclature. Besides 'sugar', there are also the terms *saccharide* and *carbohydrate*. These terms all have the same meaning. Individual sugar molecules can be linked together to form *oligo-* and *polysaccharides*. Sucrose, i.e. common table sugar, is a disaccharide; starch and cellulose are examples of polysaccharides. We will limit our discussions in this chapter to monosaccharides, because these are most relevant to nucleotide chemistry.

All monosaccharides definitionally contain one carbonyl carbon. Thus, all monosaccharides are either *polyhydroxyaldehydes* or *polyhydroxyketones*. Instead of polyhydroxyaldehyde, we can just say **aldose** and instead of polyhydroxyketone, we can just say **ketose**, where the '-ose' suffix implies sugar. Note that the terms aldose and ketose differentiate sugars based on the type of carbonyl group they contain—an aldehyde or a ketone—but we can also classify sugars based on their number of carbons. Glyceraldehyde has three carbons, so it is a triose (or more completely an aldotriose). Four-carbon sugars are tetroses, five-carbon sugars are pentoses, six-carbon sugars are hexoses, and seven-carbon sugars are heptoses, etc. The two classifications can also be combined to refer to, for example, ketohexoses or aldopentoses. Fructose is a ketohexose and ribose is an aldopentose, see Figure 5.3.

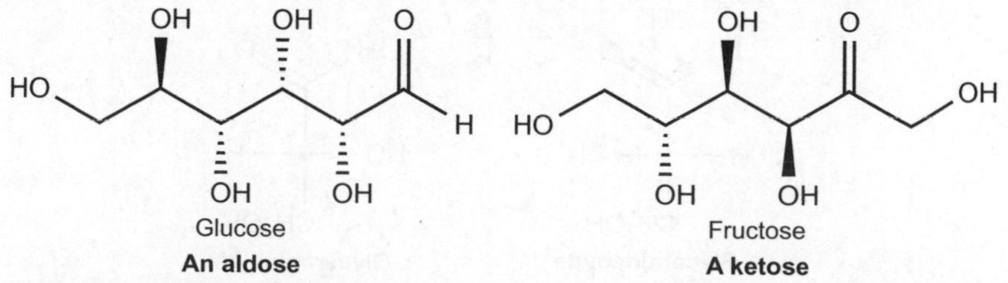

Glucose
An aldose

Fructose
A ketose

Figure 5.3 Structures of glucose and fructose in their linear forms which possess aldehyde and ketone groups, respectively.

Why is it useful to be able to refer to sugars in this manner of classification? Because as sugars get larger and larger, the total number of possible stereoisomers increases geometrically, and it becomes convenient to refer to them collectively as a common set of stereoisomers rather than trying to list each sugar's individual name. Specifically, the number of possible stereoisomers scales by the total number of asymmetric centres, n, according to the formula:

$$\text{Total stereoisomers} = 2^n$$

Glyceraldehyde has one asymmetric centre and thus has two possible stereoisomers. Aldotetroses have two asymmetric centres, and so four possible stereoisomers. Aldopentoses have three asymmetric centres, and so eight possible stereoisomers. Aldohexoses have four asymmetric centres and so 16 possible stereoisomers and so on. You can appreciate that in the context of prebiotic chemistry, synthesizing the correct sugar with the right stereochemistry is no trivial task!

Fischer projections

There is another way to draw sugars that avoids using wedged and hashed bonds to describe their stereochemistry known as Fischer projections, named after Emil Fischer (1852–1919), a Nobel-Prize winning pioneer in sugar chemistry. Fischer projections do away with wedged and hashed bonds and instead rely on only perpendicular solid lines and some assumed conventions. Let's use the simplest case of glyceraldehyde again as an example.

In a Fischer projection of a sugar, the carbonyl carbon is always drawn towards the top, and the carbon at the distal end (which in general will never be an asymmetric centre) is drawn at the bottom. The bonds to other groups/atoms on the asymmetric centres, namely H and OH, are represented with horizontal lines. By convention,

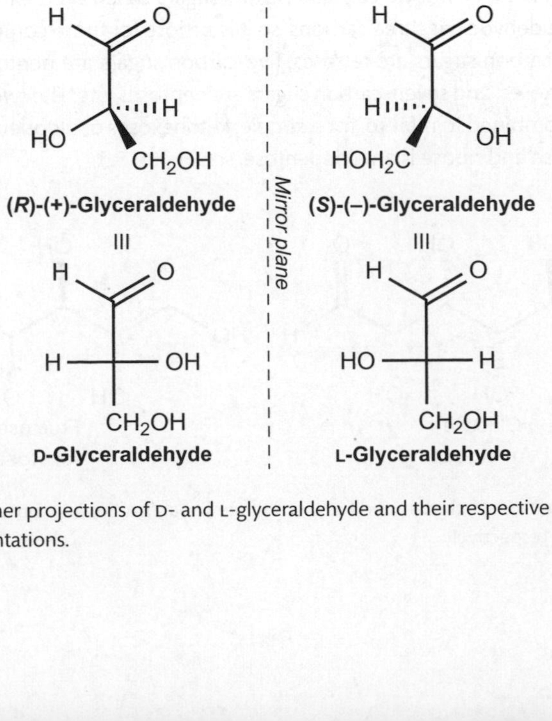

Figure 5.4 Fischer projections of D- and L-glyceraldehyde and their respective wedged/hashed representations.

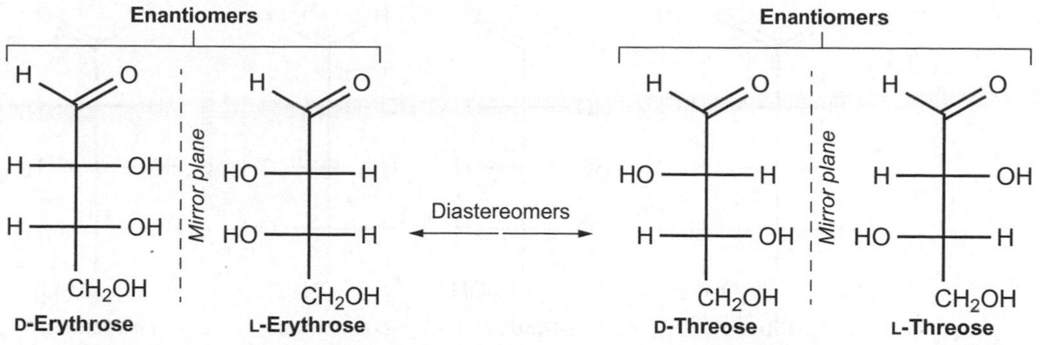

Figure 5.5 The four aldotetroses and their enantiomeric and diastereomeric relationships.

all horizontal bonds are coming towards you above the plane of the page, while all vertical bonds are going away from you below the plane of the page. Remember that glyceraldehyde has one asymmetric centre (at the second, C2 carbon), so there are two possible stereoisomers. We can draw both stereoisomers by swapping the horizontal positions of the H and OH, see Figure 5.4.

Aldotetroses have two asymmetric centres, and thus four possible stereoisomers. Notice there is a pair of stereoisomers where the OH groups are either both on the right side or both on the left, and another pair where they are on opposite sides. Each pair is a set of enantiomers, that is, they are mirror images of each other. But what is the relationship of the two pairs to each other? They have the same connectivity and molecular formula, but they can't be superimposed, and neither are they mirror images of each other. When this type of relationship is the case, we say they are **diastereomers**, see Figure 5.5.

D- and L-sugars

Finally, let's examine the aldopentoses, which have three asymmetric centres and so eight possible stereoisomers, see Figure 5.6. We'll start with the D-aldopentoses. Notice that all the OH groups on the bottom-most asymmetric centres are all on the right-hand side, while varying all the possible configurations of the top two asymmetric centres. In fact, those are *all* the D-aldopentoses. A D-sugar by definition is any sugar that when drawn in its Fischer projection has the OH group of the bottom-most asymmetric centre on the right-hand side. Also notice that each one of these D-aldopentoses are diastereomers of each other—none of them are mirror images of the other. However, we can quickly generate the remaining four of eight stereoisomers by drawing the mirror image of each of the D-isomers, and in doing so, we generate all the L-isomers, which by definition are any sugar that when drawn as its Fischer projection has the OH group of the bottom-most asymmetric centre on the left-hand side. This definition of D- and L-sugars holds for all aldoses and ketoses, and simply by drawing all the D-sugars, we can generate the L-sugars by taking their mirror images and vice versa. We could continue our discussion towards the aldohexoses and so on, but we'll stop here at the aldopentoses which are some of the most relevant sugars to prebiotic chemistry in that one in particular, D-ribose, is found in ribonucleic acid (RNA).

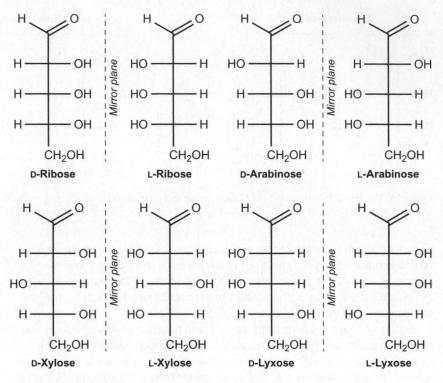

Figure 5.6 The eight aldopentoses as Fischer projections, which includes D-ribose.

Cyclic hemiacetals and anomers

So far, we've only drawn sugars in their so-called open-chain forms, but sugars can react with themselves to form cyclic structures, which generally are more thermodynamically favourable in solution. These cyclic structures result from the *intra*molecular reaction of one of the sugar's hydroxyl groups with its carbonyl carbon. Let's take a look at D-ribose as a specific example. The OH group on carbon 4 can attack carbon 1 forming a five-membered ring, or the OH group on carbon 5 can attack this same carbonyl group to form a six-membered ring. Five- and six-membered rings are especially stable and form readily. Six-membered cyclic sugars are generally referred to as **pyranoses** and five-membered rings as **furanoses**. For such cyclic sugars, instead of Fischer projections, Haworth projections are often used as shown in Figure 5.7. In aqueous solution, sugars may rapidly interconvert between open and cyclic forms, and between furanose and pyranose forms.

When an alcohol group reacts with an aldehyde in an addition reaction, the result is a hemiacetal, so we can refer to both furanoses and pyranoses as cyclic hemiacetals. But carbonyl carbons are sp^2 hybridized, meaning that it and the three bonded atoms have a trigonal planar geometry, and all lie in the same plane. The OH group can attack from either above or below the plane, resulting in a cyclic hemiacetal with a newly formed OH group either below or above the plane, respectively, with regard to the Haworth projection. Hence, two possible stereoisomers can form during cyclic

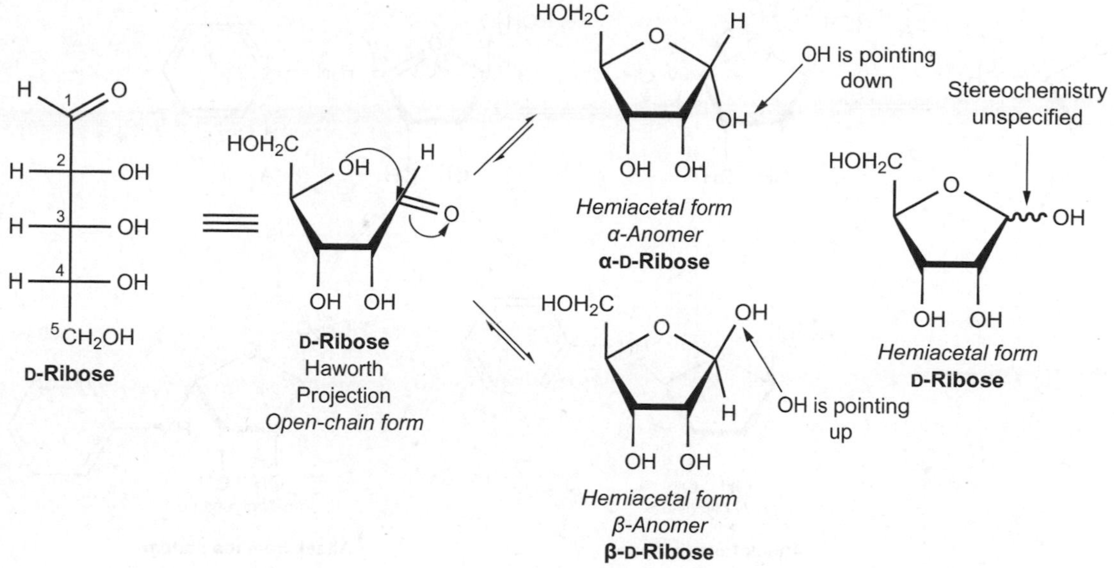

Figure 5.7 The anomeric chemistry of D-ribose when forming a furanose cyclic hemiacetal as shown in their Haworth projections.

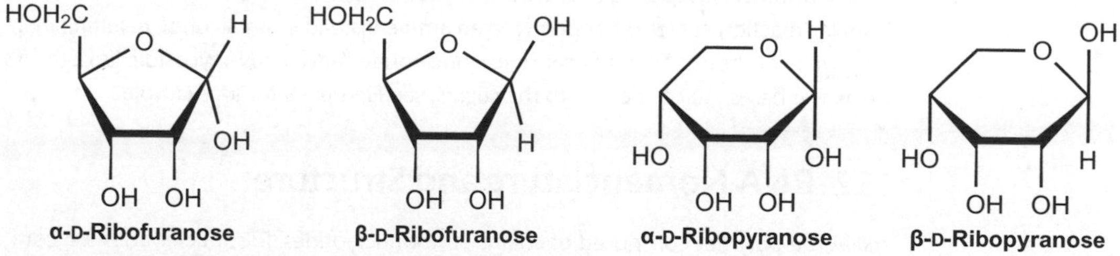

Figure 5.8 Structures for all D-ribose cyclic hemiacetals shown in their Haworth projections.

Note: That the pyranose forms are actually more stable than the furanose forms.

hemiacetal formation. These isomers are called **anomers**, and the carbon where this occurs, the anomeric carbon. *β-Anomers* have the OH group pointing up in a Haworth projection, and *α-anomers* have it pointing down. So now let's put all of our nomenclature together to give specific names to all the possible cyclic D-ribose hemiacetals, see Figure 5.8.

The β-D-ribofuranose configuration is the form found in RNA.

The glycosidic bond

We are almost done discussing basic sugar structure and nomenclature and ready to move onto nucleotides in the next section, but before we do, there is one last piece of chemistry and nomenclature that we need to discuss—the glycosidic bond. In the same way that sugars can react intramolecularly to form hemiacetals, they can react

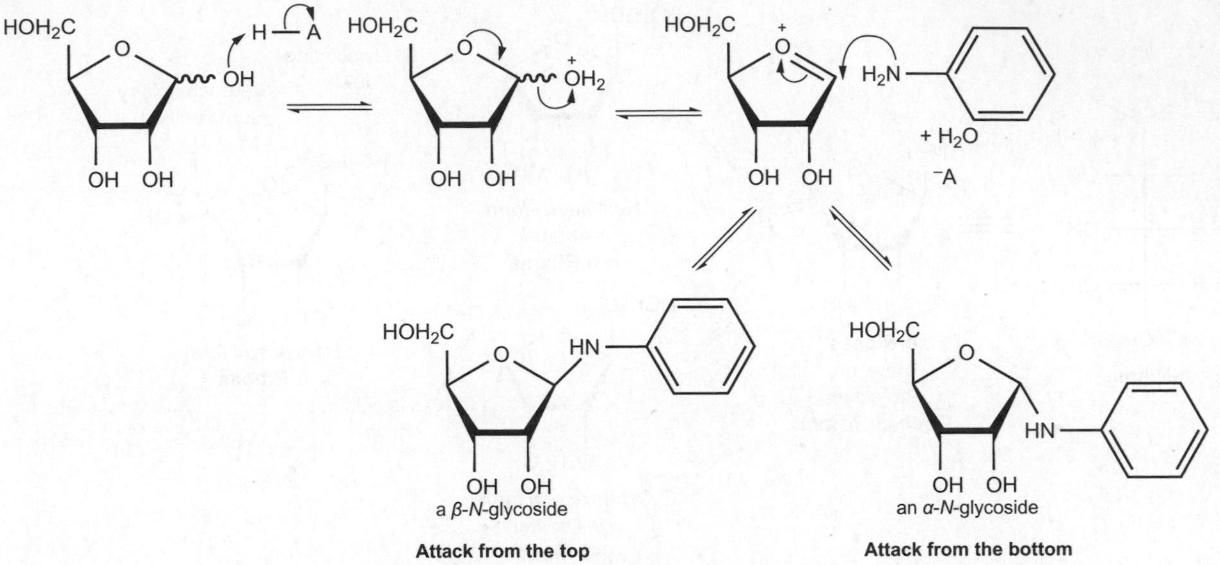

Figure 5.9 Acid-catalysed mechanism of N-glycosidic bond formation between D-ribofuranose and aniline as an example.

intermolecularly with another alcohol to form 'full-fledged' acetals. As before, both α and β anomers are possible. The new bond that results is called a glycosidic bond. A similar reaction can also happen with an amine instead of an alcohol, resulting in an N-glycosidic bond. All of the canonical nucleotides have an N-glycosidic linkage—it's how the bases are connected to the sugars, see Figure 5.9 for an example.

5.2 RNA Nomenclature and Structure

RNA is a polymer composed of chains of ribonucleotides. Ribonucleotides are composed of three structural components: a sugar, namely, β-D-ribofuranose that we have already discussed, a phosphate group, and a nucleobase, either adenine, guanine, cytosine, or uracil (A, G, C, or U). A nucleotide includes a phosphate group, whereas a **nucleoside** contains only the sugar and nucleobase components.

The nucleobases are what directly determine the genetic information or otherwise determine the catalytic activity in any given RNA sequence. Each nucleobase is either a type of substituted purine or pyrimidine, see Figure 5.10. A pyrimidine is a six-membered aromatic ring with two nitrogen atoms in the 1 and 3 positions, while a purine is a pyrimidine fused with an imidazole ring. Adenine and guanine are both purines, while cytosine, uracil, and thymine are all pyrimidines. A, G, C, and U/T nucleobases are all aromatic compounds. Having aromatic character also means each of the four nucleobases are very nearly planar (flat), a feature which is important for inter-base stacking that stabilizes the double helices often encountered in more complex RNA molecules, as well as in deoxyribonucleic acid (DNA). Their aromatic character also means they are strong absorbers of ultraviolet (UV) light in the ~240 to 300 nm region, see Figure 5.11.

As mentioned earlier, the nucleobases are connected to β-D-ribofuranose through an N-glycosidic linkage to form nucleosides. *Adenosine* is the nucleoside with the

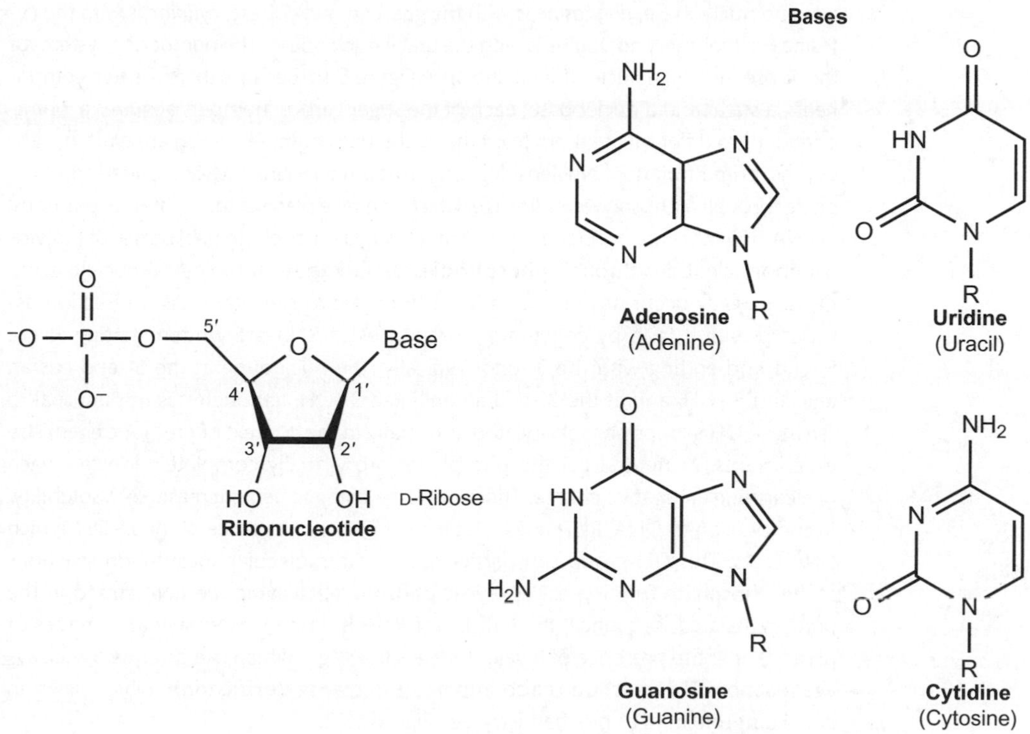

Bases

Adenosine R
(Adenine)

Uridine
(Uracil)

Guanosine
(Guanine)

Cytidine
(Cytosine)

Ribonucleotide

D-Ribose

Figure 5.10 General structure of a mononucleotide, including the ring numbering system for the ribose component (left) and the structures of the associated heterocyclic bases and names of the canonical parent nucleosides.

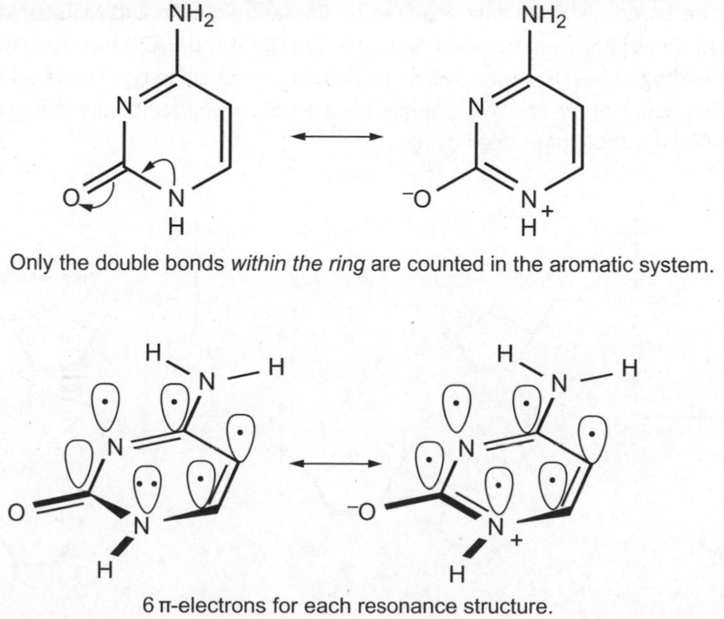

Only the double bonds *within the ring* are counted in the aromatic system.

6 π-electrons for each resonance structure.

Figure 5.11 Depiction of nucleobase aromatic resonance using cytosine as an example.

Note: All four nucleobases are aromatic structures.

adenine nucleobase, *guanosine* is with the guanine nucleobase, *cytidine* is with the cytosine nucleobase, and *uridine* is with the uracil nucleobase. The numbering system for the ribose moiety in nucleotides is shown in Figure 5.10. Because there are two components, a sugar and a nucleobase, each of the sugar carbon numbers acquires a 'prime' symbol (') to differentiate them from the nucleobase numbers which are written without the prime notation. Nucleosides are termed nucleotides when bonded to phosphate through a phosphoester linkage, which can take place at the 2′, 3′ or 5′ positions.

RNA polymers are connected through the sugar-phosphate backbones of individual ribonucleotides through **phosphodiester linkages**. In biological nucleic acids, these linkages occur at the 5′ and 3′ carbons. Moreover, RNA has a direction associated with it, and by convention, sequences of RNA are written starting at the 5′-end and ending with the 3′-end. Typically, the nucleotide at the 5′-end has an unlinked 5′-OH, and at the 3′-end an unlinked 3′-OH, although it is not unusual to see these OH groups phosphorylated, especially in the context of prebiotic chemistry experiments. At neutral pH the phosphates are virtually completely ionized, each one carrying a negative charge. These negative charges help increase RNA solubility. In comparison to DNA, RNA is less stable owing to the presence of the 2′-OH, which DNA lacks. This OH group can participate in intramolecular substitution reactions at the phosphate forming a 2′,3′-cyclic phosphate, cleaving the RNA strand in the process. As a consequence, the half-life of RNA is limited, especially at extremes of pH and/or in the presence of divalent metals like Mg^{2+}, which can catalyse this cleavage reaction. This reaction is also important in **transesterification**, which plays an important role in RNA biochemistry, see Figure 5.12.

Base-pairing acts like a 'glue' that can hold two segments of RNA together by virtue of hydrogen bonding interactions. The most common are **Watson–Crick base-pairs**, named after James Watson and Francis Crick who elucidated the DNA double helix structure based on X-ray data acquired by Rosiland Franklin and Maurice Wilkins. Watson–Crick pairs form between A/U (or A/T in DNA) and G/C, which have two and three hydrogen bonds, respectively. Their structures are shown in Figure 5.13. RNA can also adopt other forms of pairing, for example Hoogsteen pairs and guanine-uracil (G/U) wobble pairs, see Figure 5.13.

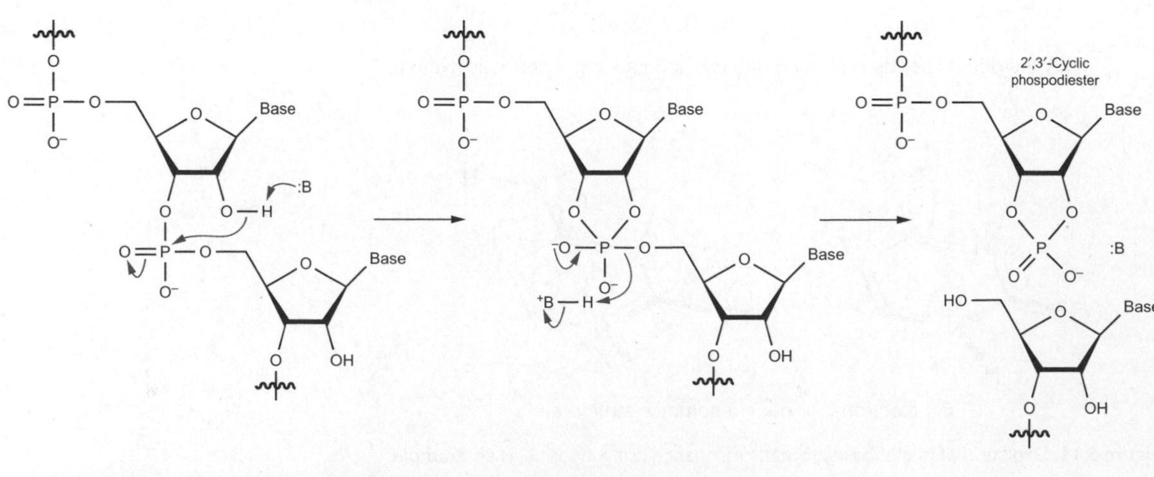

Figure 5.12 Base-catalysed RNA strand cleavage reaction/transesterification mechanism.

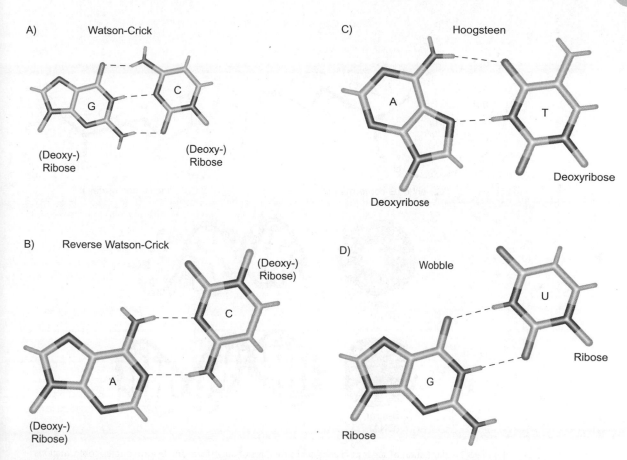

Figure 5.13 Examples of base-pairing structures including **A**) Watson-Crick, **B**) reverse Watson-Crick, **C**) Hoogsteen, and **D**) Wobble base-pairing.

Unlike DNA, RNA can take on a variety of secondary and tertiary structural motifs (see Figure 5.15), geometric features that are critical for RNA's ability to act as a catalyst as we saw for the ribosome and other naturally occurring and artificially selected ribozymes, as well as tRNA. RNA is most often found single-stranded in the cell, but it can take on a double helical structure by folding back on itself or pairing with another complementary strand. RNA forms what is called an *A-form antiparallel double helix*. This double helix structure resembles a spiral staircase with a cylindrical hole running through the centre (Figure 5.14). This structure is characterized by each of its constituent β-D-ribofuranosyl rings being present in a particular 'puckered' conformation. These rings do not lie flat as often depicted, instead the 3'-carbons are puckered up towards the nucleobases and 5'-carbons. This is called the *3'-endo* conformation. In comparison, DNA, which is often found in what is called the *B-form double helical conformation*, has its constituent deoxy-ribose units in the *2'-endo* conformation.

When a single strand of RNA folds back on itself, it often does so by forming a hairpin stem-loop structure. Other common **secondary structure** motifs include internal loops, bulges, and junctions, examples of which are shown in Figure 5.15.

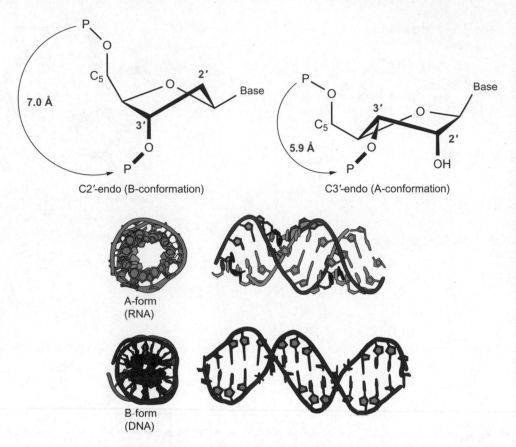

Figure 5.14 Illustration of sugar puckering and structures of the A-form and B-form duplex conformations which the different sugar puckers lead to.

Junctions form when several stem-loops meet. There are automated programs which can help predict RNA secondary structure based on sequence, for example, the UNAfold server (http://www.unafold.org/). These loops create base-pairing opportunities with other distant loops, affording **tertiary structure**. Some common tertiary structural motifs are coaxial stacking, pseudoknot formation, and ribose zippers. Coaxial stacking is often seen with junctions, for example, in tRNA.

5.3 Prebiotic Sugar Synthesis

With the basics of sugar, nucleotide and RNA structure/nomenclature discussed, we will now move onto prebiotic synthesis.

The formose reaction

Carbohydrates all share the basic elemental structure $(CH_2O)_n$. For example, glyceraldehyde has the formula $C_3H_6O_3$, and glucose has the formula $C_6H_{12}O_6$. Perhaps unsurprisingly, it was discovered in 1861 by the Russian chemist Alexander Butlerov (1828–1886), that the simplest aldehyde, formaldehyde (HCHO), can oligomerize in

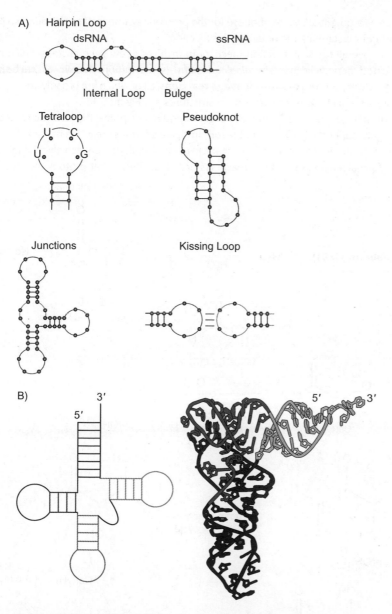

Figure 5.15 A) Examples of different types of RNA secondary structures.
B) Secondary structure of a tRNA alongside its 3D structure.

basic aqueous solution to give a wide variety of monosaccharides, including ribose. In fact, the final reaction product mixture has a smell reminiscent of maple syrup. This mixture was called 'formose' (remember the -ose suffix denotes sugar), and this reaction is now called the formose reaction, or sometimes the Butlerov reaction. In prebiotic experiments, the formose reaction is typically carried out at high pH (~10–12) in the presence of Ca^{2+} which acts as a catalyst. The formose reaction is often called upon as a potential source of ribose for the prebiotic synthesis of nucleotides, so it is worth understanding the details of this reaction.

The overall formose reaction mechanism involves a few general steps which are repeated iteratively, namely, *aldol additions*, isomerizations known as **carbonyl migrations**, and **retro-aldol cleavage reactions**, all of which are broadly applicable to sugars as a class due to their similar structures, see Figure 5.16.

It is now generally believed that a sugar is required to 'prime' the formose reaction, as highly purified HCHO does not readily instigate a formose reaction by itself. To understand why, let's examine the mechanism of aldol addition between glycolaldehyde and formaldehyde, one of the reaction's first steps. Recall that protons on carbons

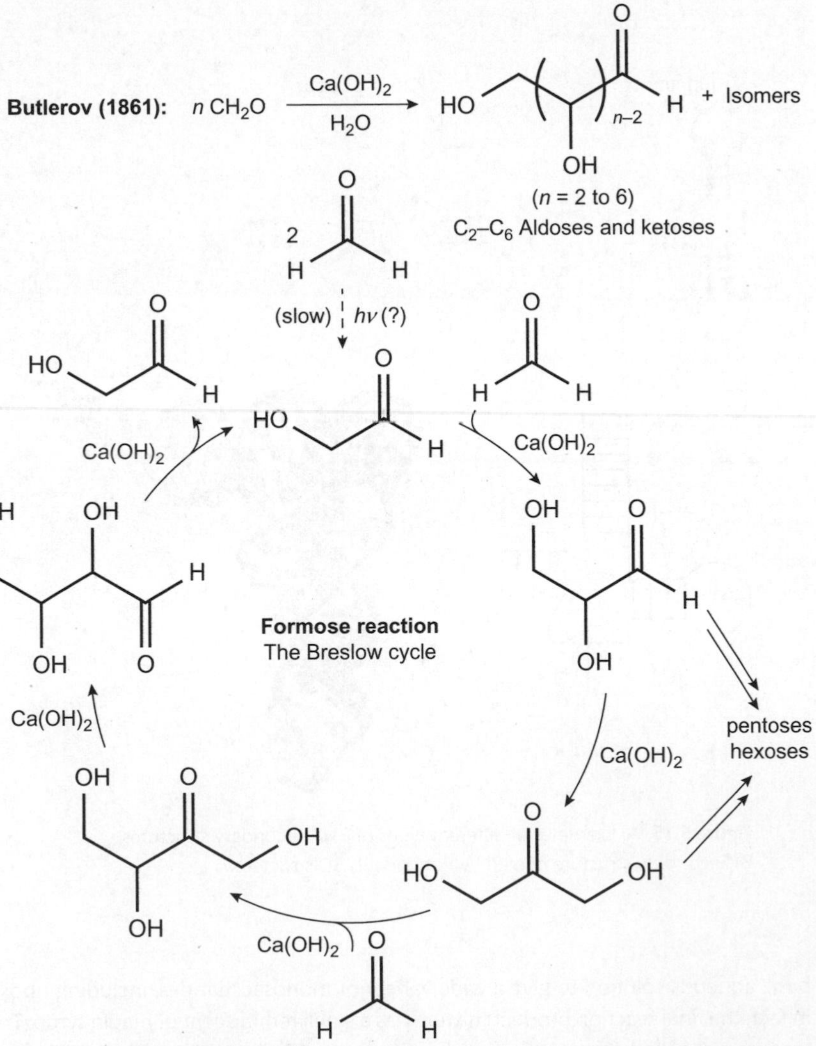

Figure 5.16 General overview of the formose reaction including the Breslow autocatalytic cycle named after Ronald Breslow (1931–2017) who proposed it.

Note: The Breslow cycle is only one possible reaction pathway in the formose reaction. The total reaction network of the formose reaction is much more complex and includes higher order sugars.

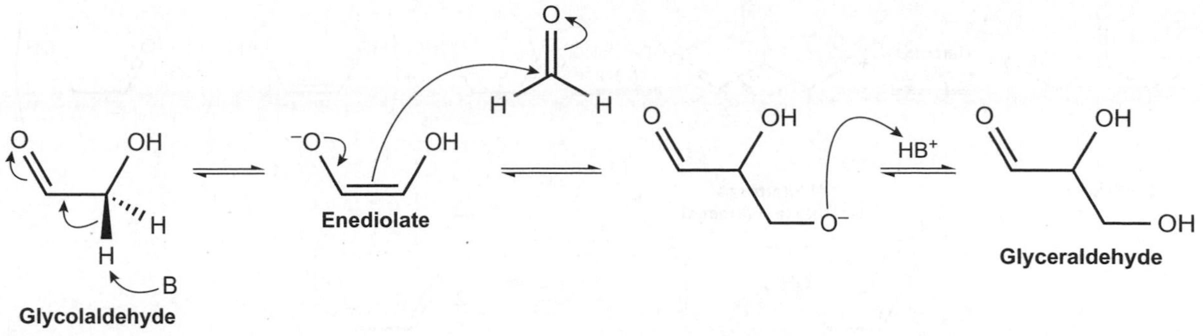

Figure 5.17 Mechanism of base-catalysed aldol addition between glycolaldehyde and formaldehyde yielding glyceraldehyde.

alpha to carbonyls are relatively acidic due to resonance stabilization afforded by the carbonyl oxygen in the conjugate base. The first step is deprotonation of an alpha proton of glycolaldehyde, which results in an *enediolate* (Figure 5.17), and after picking up a proton, an *enediol*. Enediols are good carbon-based nucleophiles (as are enediolates), with the nucleophilic atom being either carbon of the double bond, which then goes on to attack the electrophilic carbonyl carbon of formaldehyde, forming a new C–C bond. A subsequent proton transfer furnishes glyceraldehyde. Note that formaldehyde lacks an alpha carbon altogether, and so highly purified solutions of HCHO are incapable of generating the necessary enediol(ate) nucleophiles required of aldol additions. Hence, the formose reaction does not readily occur without a priming sugar. Other processes may spontaneously enable the dimerization of formaldehyde to glycolaldehyde, but these processes generally remain poorly unexplained.

The formose reaction has another fascinating feature—it is autocatalytic. To explore what this means and how it works, let's examine the next few steps in the reaction. Glyceraldehyde next undergoes isomerization via carbonyl migration to the ketone dihydroxyacetone, which becomes deprotonated at one of its alpha carbons to form a nucleophilic enediol(ate) and participates in another aldol addition reaction with formaldehyde to form the ketotetrose, erythrulose. This ketotetrose undergoes another carbonyl migration isomerization to form a mixture of aldotetrose stereoisomers (Figure 5.18). At this point, the aldotetroses can undergo retro-aldol cleavage into two molecules of glycolaldehyde, a reaction which is the reverse of aldol addition. Thus, we started with *one* molecule of glycolaldehyde and ended up with *two*. This outcome fits the definition of autocatalysis, i.e. *catalysis that occurs by one or more products of the reaction.* Since a catalyst by definition is not consumed by the reaction, you can consider one molecule of glycolaldehyde formed as the catalyst released at the end of the reaction, and the other as the product. Hence, the product of the reaction, glycolaldehyde, is also the catalyst. Autocatalysis is an important concept in prebiotic chemistry for a variety of reasons, but crucially because it represents a mechanism to achieve reproduction— you start with one copy of something, and you end up with two, see Figure 5.18.

Generally speaking, the carbonyl carbons in all sugars are electrophilic, while protons on the α-carbons are all acidic. Additionally, carbonyl migration along the backbone is generally possible for sugars. These features allow the formose reaction to produce a large mixture of higher order sugars, including branched ones. Finally, the sugars present in the reaction can undergo retroaldol fragmentation, generating a diversity of reaction products. These cycles can include four carbon or higher sugars, though the

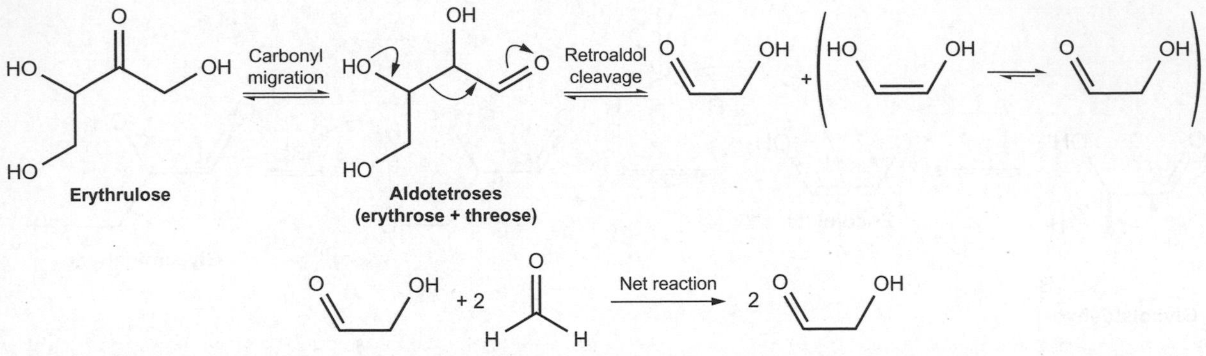

Figure 5.18 Ketotetrose carbonyl migration to aldotetroses followed by retro aldol cleavage to yield two molecules of glycolaldehyde.

Note: The net reaction of the autocatalytic cycle starting from glycolaldehyde consumes two molecules of formaldehyde to yield two molecules of glycolaldehyde.

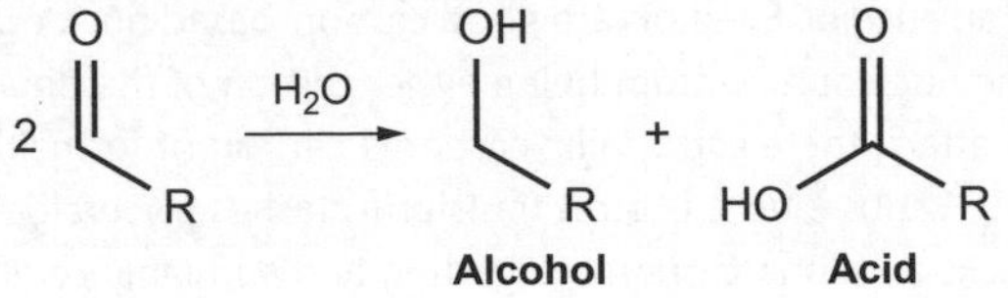

Figure 5.19 Example of the Cannizzaro reaction starting from a generic aldehyde.

reaction tends to produce a significant amount of hexoses which are able to form relatively stable nonreactive cyclic hemiacetals (pyranoses). In addition to low molecular weight carbohydrates, left to run to completion the reaction produces fairly complex higher molecular-weight 'tar-like' materials. Moreover, there is a significant amount of **Cannizzaro chemistry** proceeding in tandem, a reaction which disproportionates two aldoses to yield a carboxylic acid and an alcohol, see Figure 5.19.

A wide variety of reaction conditions and additives have now been tested with the formose reaction and many appear to be capable of steering it to favour certain reaction products. For example, the addition of borate to the reaction tends to inhibit the formation of sugars longer than pentoses by blocking sugar enediol formation. Divalent metal ions, for example Pb^{2+}, have also been shown to steer the reaction to favour particular outcomes, as has the exposure of the reaction to ultraviolet light and the presence of various mineral surfaces.

The Kiliani–Fischer synthesis

Another mechanism for making sugars is the **Kiliani–Fischer synthesis**, named after the German chemists Heinrich Kiliani (1855–1945) and Emil Fischer (1852–1919). Instead of relying on aldol additions to generate new carbon-carbon bonds, this mechanism uses cyanohydrin formation. **Cyanohydrins** result when a molecule of hydrogen cyanide (HCN) adds across a carbonyl (aldehyde or ketone), yielding a carbon bonded to both a nitrile (cyano group) and a hydroxyl (hydrin) group. In the next step, the nitrile is reduced to the imine by the formal addition of two electrons and two protons. In the last step, the imine is hydrolysed to the carbonyl. This series

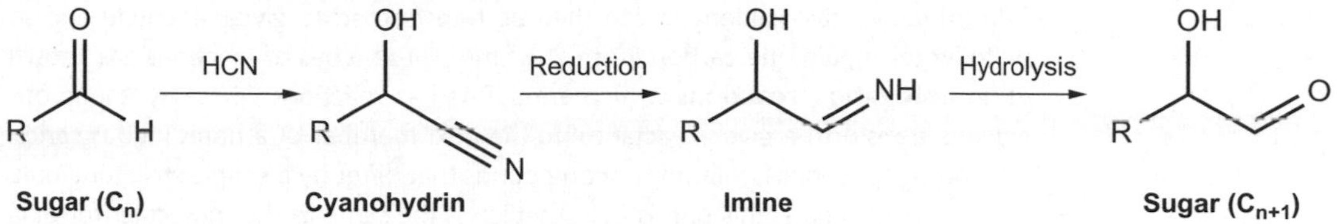

Figure 5.20 Overview of the steps involved in the Kiliani–Fischer synthesis of sugars.

of steps—cyanohydrin formation, reduction, and hydrolysis—can happen repeatedly, generating larger sugars one carbon at a time, see Figure 5.20.

Let's take a look at a specific example of a Kiliani–Fischer synthesis occurring under plausibly prebiotic conditions by starting with a mixture of formaldehyde with excess HCN in water at neutral pH. At pH 7, ~1% of the HCN is ionized as its conjugate base, $^-$CN (HCN pK_a ~ 9.2 at 25 °C). The $^-$CN anion is a good nucleophile, which can attack the electrophilic carbonyl carbon of formaldehyde resulting in an alkoxide intermediate, which quickly undergoes protonation to form the cyanohydrin known commonly as glycolonitrile. Although the formation of glycolonitrile is rapid and spontaneous, the reaction is reversible. The next step is then reduction of the nitrile, but what reducing agents exist that are also prebiotically plausible? Classic reducing agents/conditions like lithium aluminum hydride and H_2/Pd are difficult to argue as being abundant on early Earth.

One powerful reducing agent may have been quite abundant—the **hydrated electron**. As the name suggests, hydrated electrons are free electrons surrounded by a solvation shell of water. While they are short lived, hydrated electrons persist long enough to effect the reduction of the nitrile of glycolonitrile (and other cyanohydrins) to imines, proceeding through radical intermediates. Hydrated electrons can also be produced directly from water by the absorption of ionizing radiation, e.g. γ-rays or high-energy particles. The imine then undergoes hydrolysis according to the mechanism shown in Figure 5.21 to yield glycolaldehyde.

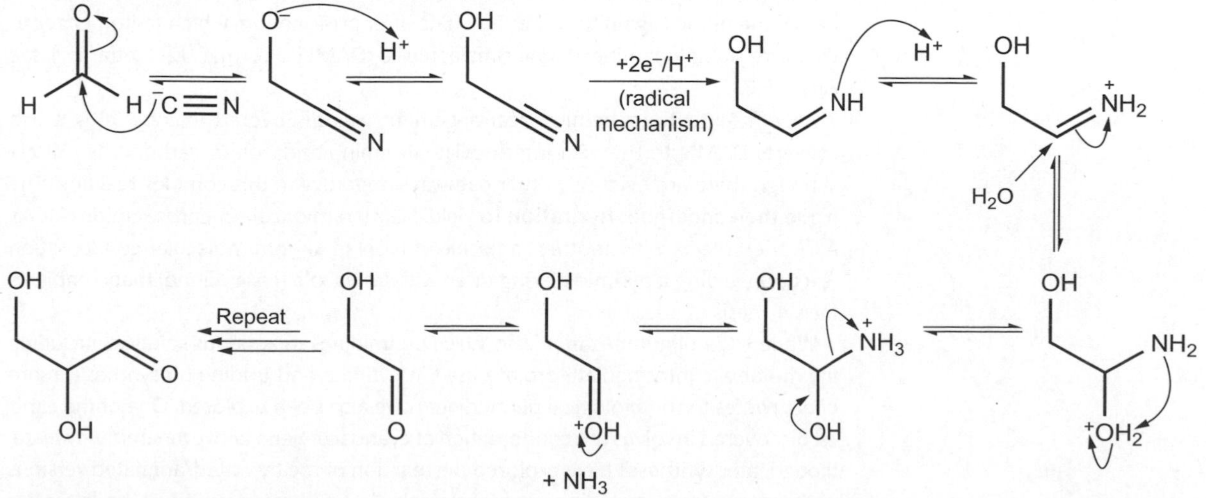

Figure 5.21 Mechanism of the Kiliani–Fischer sugar synthesis using hydrated electrons as the reducing agent. For more details, see reference 5 of the Suggested Reading.

In principle, glycolaldehyde can then be transformed to glyceraldehyde and so on to larger sugars one carbon atom at a time. These types of reactions are known as **homologation reactions**, which are defined as reactions whose synthetic operations transform a given reactant into the next member of a homologous series. A homologous series is a family of compounds that differ by a simple structural unit. In the case of sugars, this unit is CH_2O, which is the same as the molecular formula of formaldehyde. In comparison to the formose reaction, the Kiliani–Fischer synthesis is more 'orderly', generating sugars in a stepwise fashion, albeit not an autocatalytic one. Another interesting facet of the Kiliani–Fischer reaction is that it shares in common some of the same mechanistic steps involved in the Strecker synthesis of amino acids, a subject which is discussed in Chapter 6.

5.4 Nucleobase Synthesis

Not long after Miller's demonstration of the synthesis of amino acids under prebiotic conditions, scientists began to turn their attention to the possibility of other biomolecules being synthesized from plausible prebiotic precursors. HCN has long been known to be a component of cometary comae, and in 1960 Juan Oró (1923–2004) attempted the synthesis of adenine from aqueous HCN. Formally, adenine is a pentamer of HCN, $(C_5H_5N_5)$, though the mechanism of synthesis is somewhat complex. In fact, fairly concentrated HCN solutions do yield adenine among other purines (including guanine), along with a variety of other small molecules (including pyrimidines and amino acids) and a significant amount of a complex polymeric material.

HCN can act as both an electrophile, and upon deprotonation, a nucleophile. Since the pK_a of HCN is ~9.2 at 25 °C, in even slightly basic water there is a significant amount of the nucleophilic $^-$CN anion present, which is able to add to the un-ionized HCN also present in the reaction mixture. To illustrate this reactivity, we'll examine one proposed mechanism for adenine synthesis. The first step is addition of $^-$CN to the carbon of HCN forming (after protonation) formimidoyl cyanide (a formal HCN dimer). Another addition of $^-$CN to the imine carbon followed by protonation generates aminomalononitrile (AMN, a formal HCN trimer). A third addition of $^-$CN to one of the nitriles again furnishes an imine after protonation, which tautomerizes to the more stable enamine, diaminomaleonitrile (DAMN, a formal HCN tetramer), see Figure 5.22.

An efficient photochemical rearrangement was discovered in the 1960s which converts DAMN to the next intermediate, 4-aminoimidazole-5-carbonitrile (AICN), although there are evidently other pathways operative in this complex reaction. The nitrile then undergoes **hydration** to yield 4-aminoimidazole-5-carboxamide (AICA). AICA then reacts with another molecule of HCN in an intramolecular condensation reaction yielding a pyrimidine ring fused with imidazole (i.e. a purine) that completes the synthesis of adenine.

While HCN oligomerization also yields pyrimidines in aqueous solution, including the metabolic intermediate orotate used in cytidine and uridine biosynthesis, more direct routes to the biological pyrimidines have also been explored. One of the earliest discovered involves the condensation of cyanoacetylene with cyanamide/cyanate, though later syntheses have explored the reaction of the hydrated/aminated versions of these reactants, including cyanoacetaldehyde and urea/guanidine. The latter has especially been explored under so-called 'drying conditions' that might be expected to occur on the edge of primitive drying puddles or lagoons, see Figure 5.23.

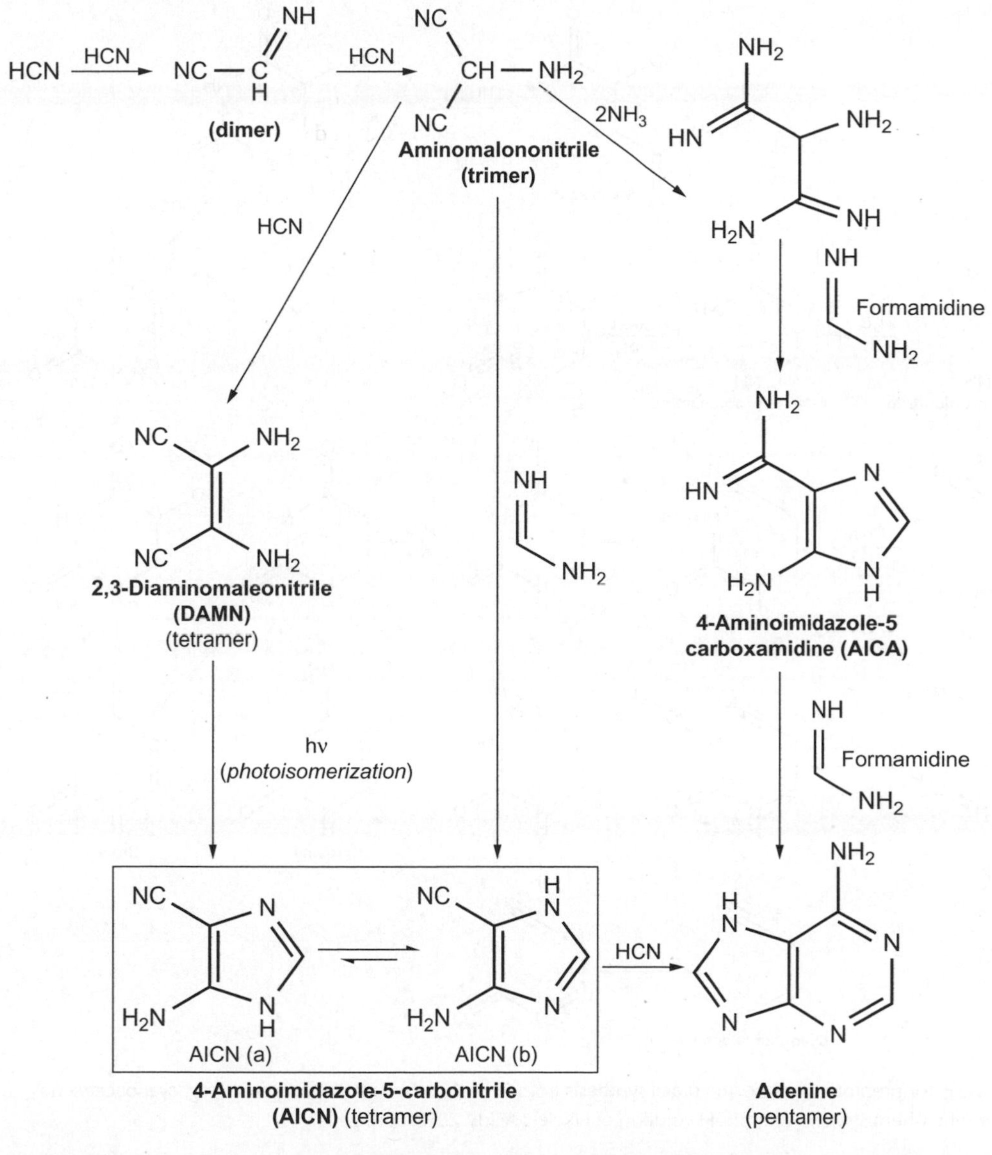

Figure 5.22 Pathways for the prebiotic synthesis of adenine from HCN. Scheme adapted from D. Roy et al. (2007). 'Chemical Evolution: The Mechanism of the Formation of Adenine under Prebiotic Conditions'. *PNAS* *104*: 17272–17277.

While in biochemistry, cytidine derivatives are typically derived from amination of uridine derivatives, the opposite is also possible: uracil derivatives can be derived via hydrolytic deamination of cytosine derivatives. Thus, it is widely considered that synthetic conditions that allow for the production of cytosine or cytidine also allow for the synthesis of uracil or uridine derivatives.

While the heterocycle syntheses presented here offer a few different approaches to forming such molecules from simple precursors in water, there may of course be others, and the environment may have supplied molecules besides those present in contemporary biochemistry.

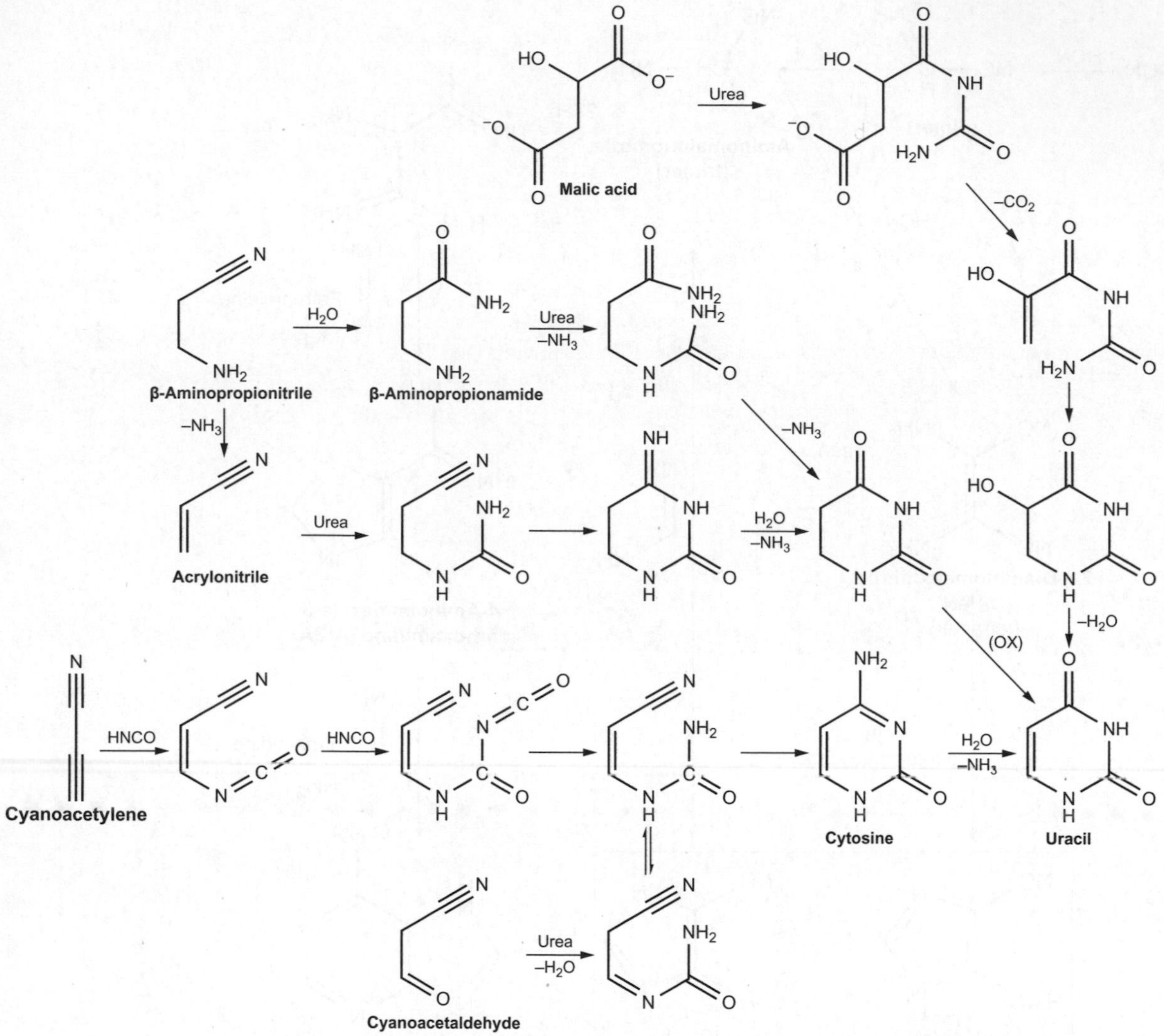

Figure 5.23 Different pathways for prebiotic cytosine and uracil synthesis from 3- or 4-carbon starting materials, e.g. cyanoacetylene. Scheme adapted from 'Prebiotic Chemistry and Chemical Evolution of Nucleic Acids'. 2018. Springer.

5.5 Ribonucleoside/tide Synthesis

Of course, nucleobases are only one step along the way to the components of RNA; if RNA was important for the origins of life, then there would need to be abiotic mechanisms for converting the nucleobases into ribonucleosides/tides. Alternatively, one can imagine syntheses in which the entire ribonucleosides/tides including the nucleobase are constructed in tandem. Modern organic chemistry recognizes that there are many synthetic routes to the same synthesis target; the one chemists choose is often determined by factors including speed, ease, and cost, and these considerations generally reflect the state of the chemical art. Biosynthesis, in contrast, is governed by the evolution-dependent discovery of enzymatic reaction mechanisms to produce compounds which are of unknown utility to chemical systems, and then selected according to their contribution to the survival of the system. Thus, there may be little convergence between what pathways were available to prebiotic systems and the ones biology selected. One might conceptualize this as being like a braided river delta. The river flows downhill but may follow many paths to reach the same final outcome. That does not mean the various potential flow paths are unimportant, but that any description of them may not be exhaustive. Indeed, the RNA world model sometimes makes the assumption that RNA must be the outcome, and the preferred outcome, of chemical and biological evolution, because it exists and underlies all terrestrial biology.

One of the first attempts at a prebiotic synthesis of ribonucleosides involved simply drying purines with ribose in a sort of 'forced' condensation reaction driven by evaporation of H_2O. The formation of an N-glycosidic bond between ribose and a nucleobase requires the substitution of the OH group on the anomeric carbon along with removal of a proton on the nitrogen, which together generate a molecule of water. The result of this reaction is a mixed yield of α and β anomeric ribonucleosides as well as exocyclic-amine glycosylated derivatives, see Figure 5.24A. Pyrimidines tend not to react as robustly via this drying mechanism, presumably because the pyrimidine rings tend to be poorer nucleophiles. In fact, the lone pair of electrons on the pyrimidine N1 nitrogen is part of the aromatic π system and not available for nucleophilic attack. In order to make N1 nucleophilic, either deprotonation or **tautomerization** to the enol form must occur for both cytosine and uracil, see Figure 5.24B.

In response to this lack of reactivity of the pyrimidines with ribose using drying conditions, the strategy of constructing the nucleobase in tandem with the sugar was developed. Rather than synthesizing the ribose sugar outright, a bicyclic intermediate known as D-*arabinofuranosyl aminooxazoline*, is synthesized first over a couple of steps from glycolaldehyde, D-glyceraldehyde and cyanamide. (Recall that glycolaldehyde and glyceraldehyde can both be generated stepwise by Kiliani–Fischer chemistry.) Note that the D-arabinofuranose ring is a five-membered pentose hemiacetal, which differs from the D-ribose hemiacetal by only the stereochemical configuration at the 2-carbon. First, glycolaldehyde reacts with cyanamide via cyclization to generate a 2-aminooxazole ring. This compound has a nucleophilic carbon (reminiscent of an enediol), which then attacks the electrophilic carbonyl carbon of D-glyceraldehyde and then cyclizes to furnish the D-arabinofuranosyl aminooxazoline, along with other stereochemical and pyrano-isomers. D-Arabinofuranosyl aminooxazoline can then react with cyanoacetylene (which itself is a product of spark-discharge acting upon gaseous mixtures of methane and nitrogen) in another cyclization reaction, to yield 2,2'-anhydrocytidine. This compound can react with phosphate to yield β-ribocytidine-2',3'-cyclic phosphate among other phosphorylated products. The cytosine nucleobase can then be deaminated by hydrolysis at the 4-carbon to yield the corresponding uridine nucleotide, see Figure 5.25.

While this strategy for the prebiotic synthesis of the pyrimidine ribonucleotides is ingenious, it is quite different from how ribonucleotides are biosynthesized. If this were the way ribonucleotides originated on early Earth, then it would have to mean that biology, over time, completely reinvented the process. Recall from Chapter 4 that the biosynthetic pathways for both purine and pyrimidine ribonucleotides utilize preformed 5-phosphoribosyl pyrophosphate. Biological nucleobases are not constructed from nitriles, but rather amino acids, glutamine, bicarbonate, and formyl

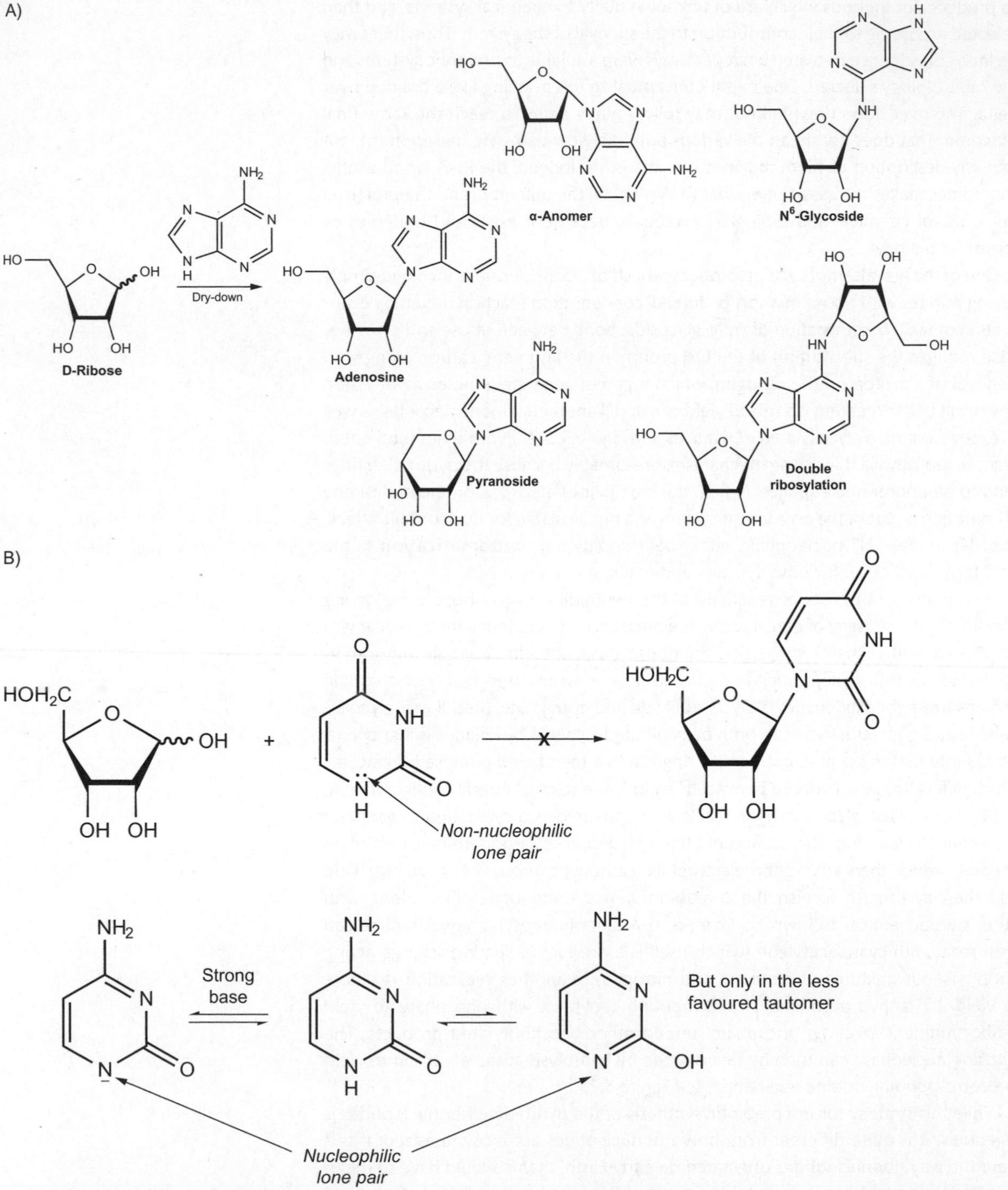

A)

B)

Figure 5.24 A) Nucleoside products produced by the dry-down reaction of adenine with D-ribose (shown in its furanose form). Note that the canonical adenosine structure is only one of the nucleoside products formed in this reaction. **B)** A similar reaction using pyrimidine nucleobases does not produce nucleoside products. This is likely the result of the non-nucleophilic lone pair, which can only be made nucleophilic by strong base or tautomerization. For more details, see W. D. Fuller et al. (1972). 'Studies in Prebiotic Synthesis. VII: Solid-State Synthesis of Purine Nucleosides'. *J Mol Evol* 1: 249–257.

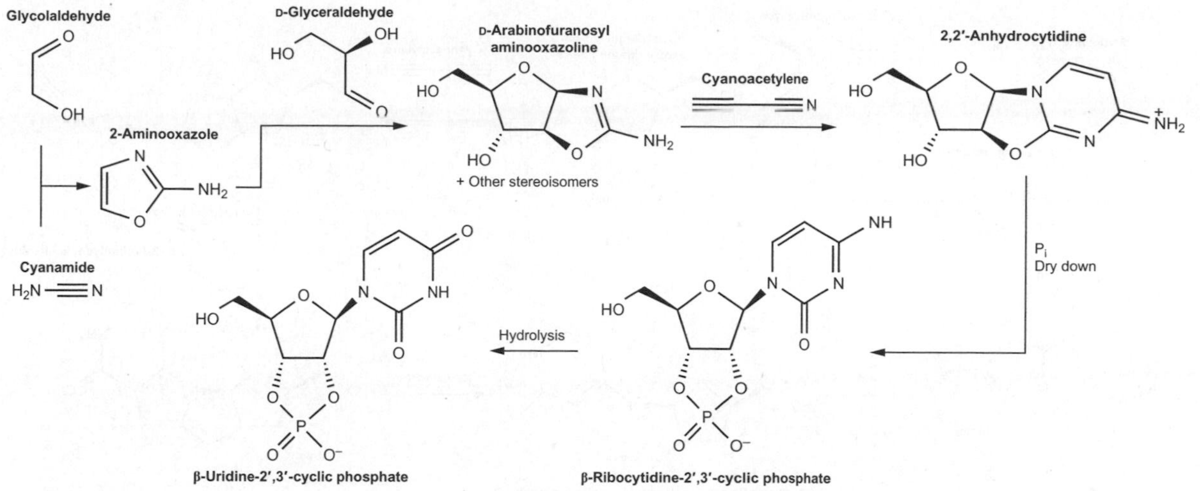

Figure 5.25 Prebiotic pathway for pyrimidine nucleotide synthesis via the D-arabinofuranosyl aminooxazoline intermediate. *Note:* This synthesis is often referred to as the Powner-Sutherland pathway, see reference 5 in the Suggested Reading for more details.

groups donated by the co-enzyme formyl-tetrahydrofolate (THF). Although it's difficult to rule out the possibility that biology completely reinvented these pathways, there are some aspects of investigated prebiotic ribonucleoside/tide syntheses that do resemble some features and steps of extant biosynthetic pathways. This may not mean that these pathways' overlaps with extant biosynthesis are ontologically meaningful, and there of course remain an untold number of combinations of reactions and reaction conditions that could lead to biological ribonucleotides, or alternative ones, as products.

Purine ribonucleotides, at least as their 8-oxo derivatives, can also be synthesized making use of a similar D-arabinofuranosyl aminooxazoline intermediate albeit with a thiol in place of the amine, see Figure 5.26. In the first step, rather than reacting with cyanamide, its molecular cousin thiocyanic acid is reacted with glycolaldehyde to yield 2-thiooxazole. This compound is cyclized again with D-glyceraldehyde like before, yielding a mixture of pentose oxazolidinone thiones, including the arabino derivative. The sulfur atom then reacts as a nucleophile with the terminal carbon of cyanoacetylene, but in this instance, further cyclization does not happen. Instead, the resulting thiovinylnitrile serves as a leaving group, which can be substituted by electron-deficient amines, for example, 2-aminocyanoacetamide. The nitrogen atom of the oxazoline subunit then attacks the nitrile carbon in a cyclization reaction. Removal of a C–H proton by base results in a fused imidazole ring. (Recall in purine biosynthesis that the imidazole ring is synthesized first, followed by the pyrimidine ring, both by intramolecular condensation reactions involving a formyl group.) Next, formamidine is introduced, which reacts at the primary amine forming an imine, which goes on to condense with the neighbouring amide group in a cyclization reaction forming the fused pyrimidine ring. As before, reaction of the resulting anhydronucleoside with inorganic phosphate yields the 2',3'-cyclic phosphate with the concomitant formation of the β-ribofuranosyl N-glycosidic linkage and an 8-oxohypoxanthine nucleobase. This prebiotic pathway resembles the purine biosynthetic pathway in several ways, namely, that the imidazole ring is constructed first followed by the fused pyrimidine ring, and the fused pyrimidine ring-forming

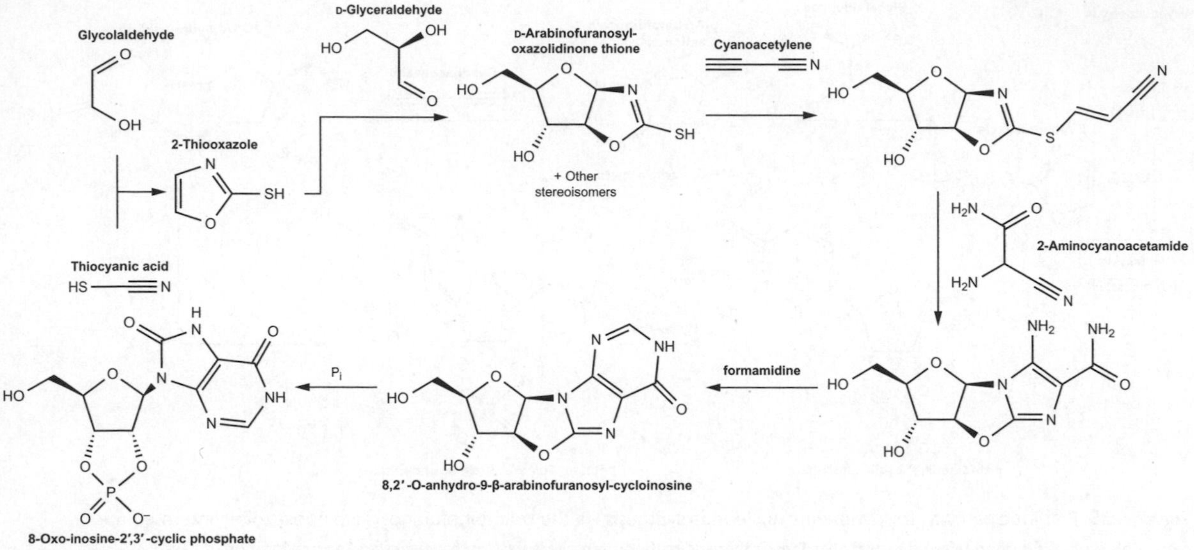

Figure 5.26 Prebiotic 8-oxopurine ribonucleotide synthesis via the oxazolidinone thione intermediate. For more details, see S. Stairs et al. (2017). 'Divergent Prebiotic Synthesis of Pyrimidine and 8-Oxo-Purine Ribonucleotides'. *Nature Commun.* 8: 15270.

step involves a condensation reaction with what effectively is a formyl group derivative, see Figure 5.26.

The following is another example of a prebiotic ribonucleotide synthetic pathway that can be compared to biology (Figure 5.27). This pathway involves the reaction of ribose-1,2-cyclic phosphate, a starting material not too different from 5-phosphoribosyl pyrophosphate, with different purine and pyrimidine derivatives that can directly furnish β-*N*-glycosidic linkages. The reaction is stereospecific in that only the β-anomer is formed—the cyclic structure ensures that attack can only take place on the β-face and not the α-face. This attack substitutes the phosphate on the anomeric carbon, yielding 2′-ribonucleoside monophosphates. This synthetic strategy resembles the biosynthetic strategy for the pyrimidine nucleotides, which also involves *N*-glycosidic bond formation of 5-phosphoribosyl pyrophosphate via a substitution with orotate, a preformed pyrimidine derivative. It also generally resembles the salvage pathways for nucleotide assembly. Purine ribonucleotide biosynthesis, on the other hand, employs the strategy of constructing the fused pyrimidine-imidazole hypoxanthine ring directly on the 1-position of 5-phosphoribosyl pyrophosphate. In a sense, biology first goes about coupling an 'unfinished' nucleobase to D-ribose, and then goes on to complete the synthesis.

Another prebiotic synthesis of purine ribonucleotides has been shown to follow a similar strategy. In this pathway, however, the pyrimidine portion of the ring is constructed first, namely 4,5,6-triaminopyrimidine. The 5-amino group is then formylated, yielding a formamido group in this position, which plays an important role in the subsequent ring-forming step. This 4,6-diamino-5-formamidopyrimidine is then reacted with D-ribose in the dry state to produce an *N*-glycosidic linkage at the 4-position of the pyrimidine derivative. The nitrogen atom of this glycosidic bond then goes on to attack the neighbouring formyl (carbonyl) carbon of the formamido group in an intramolecular condensation reaction that yields the adenine ring with an *N*-glycosidic bond to D-ribose. This step also resembles the purine biosynthesis

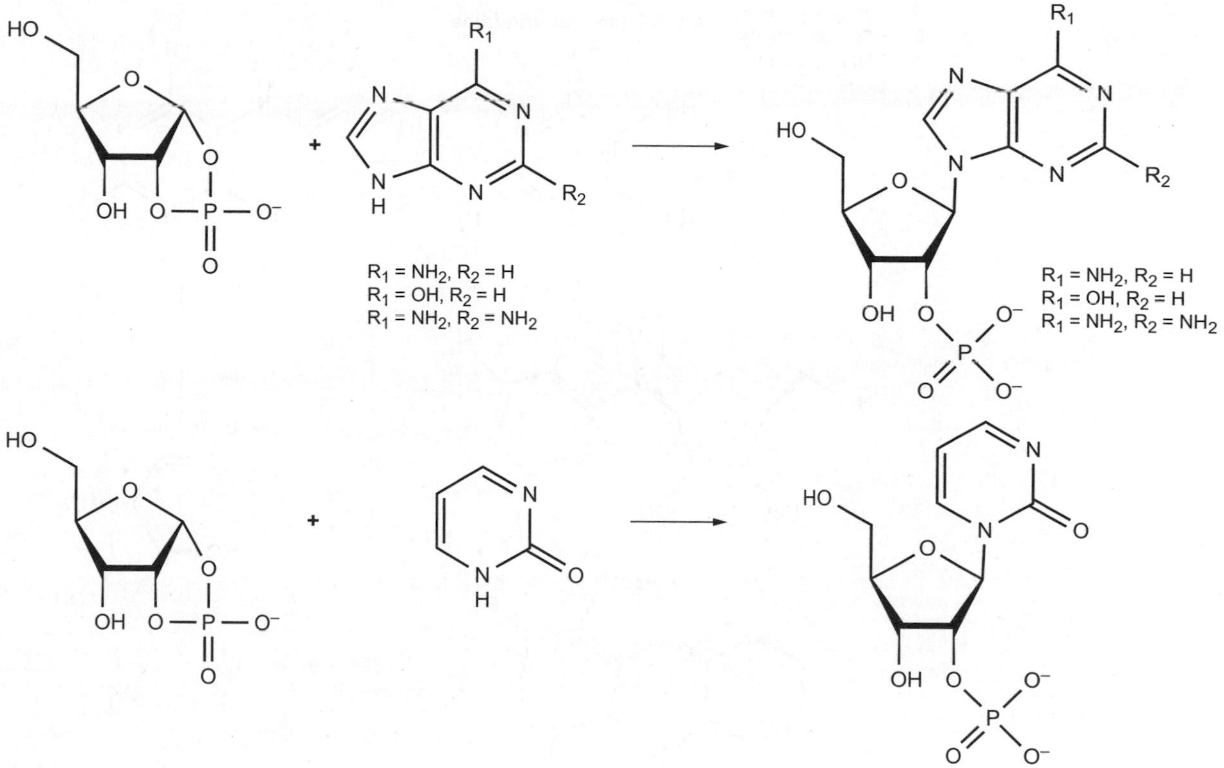

Figure 5.27 Purine and pyrimidine prebiotic ribonucleotide synthesis via direct *N*-glycosidic bond formation with ribose-1,2-cyclic phosphate. For more details, see H.-J. Kim, S. A. Benner. (2017). 'Prebiotic Stereoselective Synthesis of Purine and Noncanonical Pyrimidine Nucleotide from Nucleobases and Phosphorylated Carbohydrates'. *PNAS 114*: 11315–11320.

pathway. The glycosidic bond-forming step does, however, produce a mixture of α and β anomers along with both furanosyl and pyranosyl isomers. Nevertheless, the biological adenosine ribonucleoside with the β-furanosyl ring is among them, see Figure 5.28.

A similar strategy has also been shown for the pyrimidines, where D-ribose is first coupled with an 'unfinished' nucleobase followed by its completion. In this example *N*-isoxazolyl-urea is reacted with D-ribose to form the *N*-glycosidic linkage. The construction of the pyrimidine ring is then completed in a subsequent reduction-followed-by-cyclization reaction step yielding cytidine and uridine. As before, a mixture of α and β as well as furanosyl and pyranosyl isomers are formed, see Figure 5.29.

While it is not possible to cover all potentially prebiotic reported syntheses of nucleosides/tides and their sugar/nucleobase components, we refer the interested student to the literature at the end of this chapter.

Phosphorylation

While some of the aforementioned syntheses produce ribonucleotides directly, the direct phosphorylation of ribosides has also been explored using possibly prebiotic sources of phosphate.

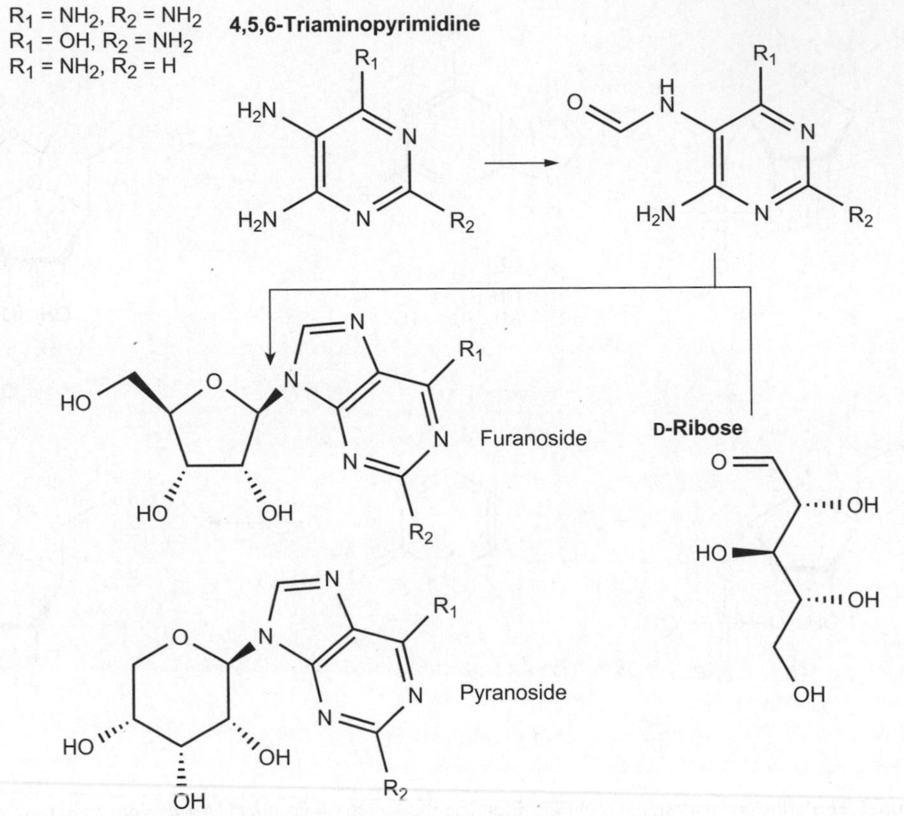

$R_1 = NH_2, R_2 = NH_2$
$R_1 = OH, R_2 = NH_2$
$R_1 = NH_2, R_2 = H$

4,5,6-Triaminopyrimidine

Furanoside

D-Ribose

Pyranoside

Figure 5.28 Purine ribonucleoside synthesis starting from 4,5,6-triaminopyrimidines. For more details, see S. Becker et al. (2016). 'A High-Yielding, Strictly Regioselective Prebiotic Purine Nucleoside Formation Pathway'. *Science* 352: 833–836.

While phosphorus is not an especially rare element in Earth's crust, neither is it especially abundant, being typically present in many rocks in the form of traces of extremely insoluble phosphate minerals including the **apatite** series ($Ca_5(PO_4)_3X$, where X can be F, Cl or OH). The P in phosphate (PO_4) is in the +V oxidation state; the lower oxidation states of P are typically not stable in environments in which liquid water is plausibly stable. Furthermore, phosphate is the conjugate base of a triprotic acid (phosphoric acid, H_3PO_4) with pK_a values of 2.1, 7.2, and 12.3, and readily forms insoluble complexes with Mg^{2+} and Ca^{2+}.

Apatites are weakly reactive with organic alcohols under drying and heating conditions, affording small amounts of phosphate esters. (Phosphate esters require energy for their formation; in general their synthesis requires of the order of 3.3 kcal mol^{-1}. This energy can be input through dry-down or chemical activation.) This being the case, prebiotic chemists have explored various alternative sources of P and ways of rendering phosphate more soluble. A simple way of making phosphate more soluble is by lowering the pH—under extremely acidic conditions apatite dissolves readily (since our teeth are mainly apatite, this is one good reason to avoid soda pop). Another way is by converting it to more soluble phosphate minerals, for

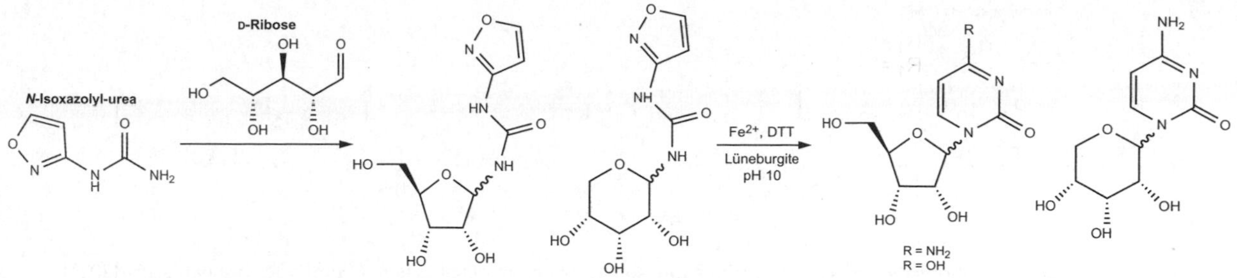

Figure 5.29 Prebiotic pyrimidine ribonucleoside synthesis starting from *N*-isoxazolyl-urea. For more details, see S. Becker et al. (2019). 'Unified Prebiotically Plausible Synthesis of Pyrimidine and Purine RNA Ribonucleotides'. *Science 366*: 76–82.

example struvite ($NH_4MgPO_4 \cdot 6H_2O$) which can form in environments containing large amounts of ammonia.

Another phosphorylation method involves the use of potentially prebiotic condensing agents such as cyanate and urea. These reactions typically afford phosphate esters in a few percent yield.

The conversion of nucleosides to nucleotides is of prime importance for RNA world models, but nucleosides have multiple modifiable hydroxyl groups (e.g. 2′, 3′, and 5′), thus a variety of products is often possible, including multiply phosphorylated and cyclic phosphorylated species, see Figure 5.30.

Yet another possible route to prebiotic phosphorylation is through condensed phosphates. In general, water-eliminating condensation reactions can be driven by heating to dryness at temperatures above the boiling point of water or in low water activity solvents such as formamide (which has often been argued to be prebiotically plausible as a relatively stable hydration product of HCN), and condensed phosphates such as pyrophosphate and trimetaphosphate (TMP) have been detected in volcanic fumaroles, presumably derived from thermal condensation. The conversion of a condensed phosphate to a phosphate ester is considerably more thermodynamically favourable than the direct condensation of orthophosphate to form a

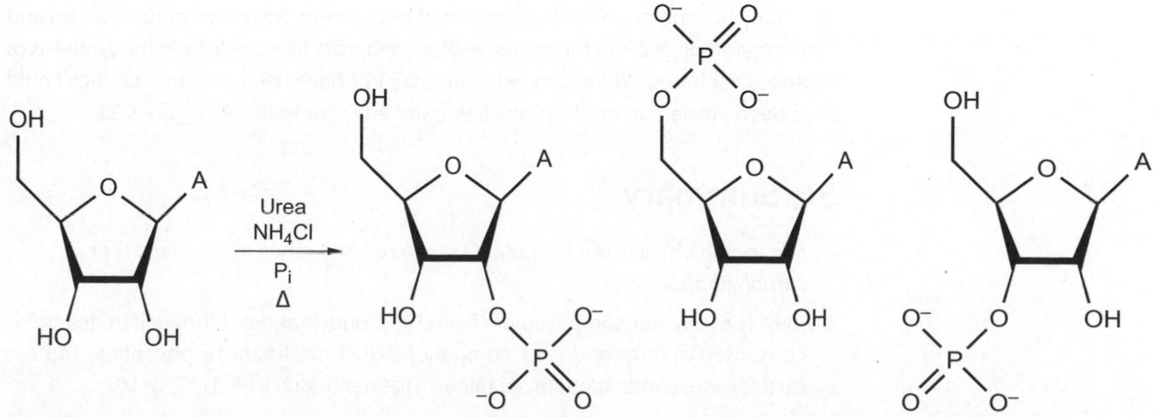

Figure 5.30 Outcome of a model phosphorylation using adenosine yielding a mixture of phosphorylated products.

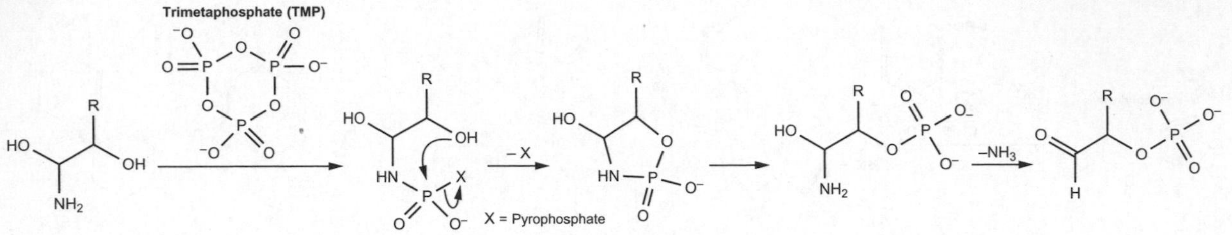

Figure 5.31 Prebiotic phosphorylation mediated by trimetaphosphate. For more details, see A. Osumah, R. Krishnamurthy. (2021). 'Diamidophosphate (DAP): A Plausible Prebiotic Phosphorylating Reagent with a Chem to BioChem Potential?'. *ChemBioChem 22*: 3001–3009.

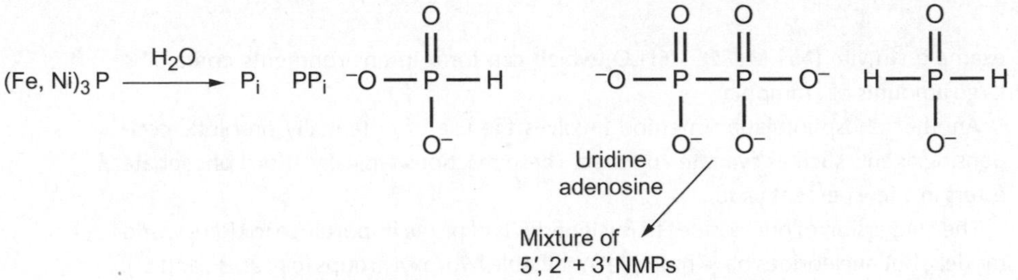

Figure 5.32 Schreibersite-mediated prebiotic phosphorylation of uridine and adenosine ribonucleosides.

phosphate ester. Thus, it has been shown that TMP will spontaneously phosphorylate nucleosides, see Figure 5.31. Various amines may catalyse the reaction, and amidated triphosphate derivatives are kinetically more favourable reagents.

Finally, while the lower oxidation states of P are not stable under conditions which allow for water to be thermodynamically stable, it is likely that significant amounts of metal phosphides, for example the mineral schreibersite ((Fe,Ni)$_3$P) which formed in the protosolar nebula reached the Earth's surface early in its history. Indeed, this mineral makes up a significant fraction of the composition of iron-nickel meteorites found on Earth, and when this mineral is placed in water it rapidly corrodes to produce higher oxidation state phosphorus compounds including phosphides and phosphonates, and ultimately phosphate, including condensed phosphates. Reactions of nucleosides and other organic alcohols in the presence of schreibersite likewise lead to the synthesis of organophosphates. While such reactions may not have been widespread, they could have been efficient in localized shallow water environments, see Figure 5.32.

5.6 Summary

- Sugars, which have the formula $(CH_2O)_n$ are also known as *saccharides* or *carbohydrates*.

- RNA is a polymer composed of chains of ribonucleotides. Ribonucleotides are composed of three structural components: β-D-ribofuranose, phosphate and a nucleobase, either adenine, guanine, cytosine, or uracil (A, G, C, or U).

- Sugars can be made prebiotically by the formose reaction and the Kiliani–Fischer synthesis.

- Adenine and other purines and pyrimidines can be produced starting from aqueous HCN. Various prebiotic syntheses for the biological pyrimidines have also been explored.

- Prebiotic syntheses of ribonucleotides are different from how ribonucleotides are biosynthesized. If this were the way ribonucleotides originated on early Earth, then biology, over time, completely reinvented the process.

5.7 Exercises

1. Can D-ribose also form a 6-membered cyclic hemiacetal? If so, draw the structure of both anomers in their Haworth projections.

2. Aldol addition is a reversible reaction, wherein the back reaction is called retro aldol cleavage. Draw the base-catalysed mechanism of retroaldol cleavage of 3-hydroxybutanal.

3. If we start with glycolaldehyde and racemic glyceraldehyde, how many possible *aldose* stereoisomers will be formed from their aldol condensation? Are aldoses the only type of sugar that can be formed from this reaction?

4. In Figure 5.25, D-glyceraldehyde is shown to react with 2-aminooxazole to produce D-arabinofuranosyl aminooxazoline. Draw the other furano-stereoisomers that are produced from this reaction using wedged and hashed bonds.

5. Do you think the hydrogen bonding involved in an A-U Watson–Crick pair would increase uracil's aromatic stability? Draw resonance structures to justify your answer.

5.8 Suggested Reading

1. H. James Cleaves II. (2008). 'The Prebiotic Geochemistry of Formaldehyde'. *Precambrian Res 16*(3–4): 111–118.

2. Robert Shapiro. (1988). 'Prebiotic Ribose Synthesis: A Critical Analysis'. *Origins Life Evol Biosphere 18*: 71–85.

3. Mahipal Yadav, Ravi Kumar, and Ramanarayanan Krishnamurthy. (2020). 'Chemistry of Abiotic Nucleotide Synthesis'. *Chem Rev 120*(11): 4766–4805.

4. Steven A. Benner, Hyo-Joong Kim, and Matthew A. Carrigan. (2012). 'Asphalt, Water, and the Prebiotic Synthesis of Ribose, Ribonucleosides, and RNA'. *Acc Chem Res 45*(12): 2025–2034.

5. John D. Sutherland. (2016). 'The Origin of Life—Out of the Blue'. *Angew Chem Int Ed 55*(1): 104–121.

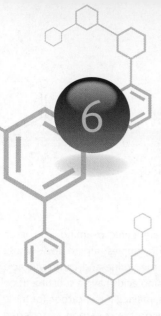

6 Aqueous Phase Amino Acid Chemistry

We continue our dive into organic chemistry with a description of prebiotic amino acid chemistry. First, we will discuss amino acid structure and nomenclature, including the difference between D and L enantiomers (building on the stereochemical discussion introduced in Chapter 5), sidechain properties, and types of biological and nonbiological amino acids. The fact that modern life almost exclusively uses the L enantiomers of amino acids will be connected to the unexplained origin of homochirality. Examples of prebiotic amino acid synthesis via the Strecker mechanism will be discussed. Other mechanisms will also be introduced. Amino acid synthesis in the context of the classic Miller–Urey and related experiments will be described. The amino acids which have been found in meteorites and some models for their synthesis will also be examined.

6.1 Amino Acid Structure and Nomenclature

Besides nucleotides, amino acids are probably one of the most important and well-studied classes of molecules in prebiotic chemistry. As demonstrated by Miller's classic spark-discharge experiment discussed in Chapter 2, they are relatively easy to synthesize in comparison to the nucleotides and are stable enough to accumulate over moderately long timescales in contrast to free sugars which decompose somewhat rapidly even under clement conditions. Amino acids are also the building blocks of proteins, and for this reason alone need to be understood in some detail. In this section we will start with the fundamentals of amino acid nomenclature and structure.

α-Amino acids

Amino acids by definition contain an amino group and a carboxylic acid group. When the amino group is bonded to the carbon adjacent (α) to the carboxylic acid, we call them α-amino acids. Proteinogenic amino acids, i.e. those which are used to make biological proteins, are all α-amino acids. When the amino group is bonded to the carbon beta to the carboxylic acid, we call it a β-amino acid, and when bonded to the gamma carbon, a γ-amino acid, and so on, see Figure 6.1.

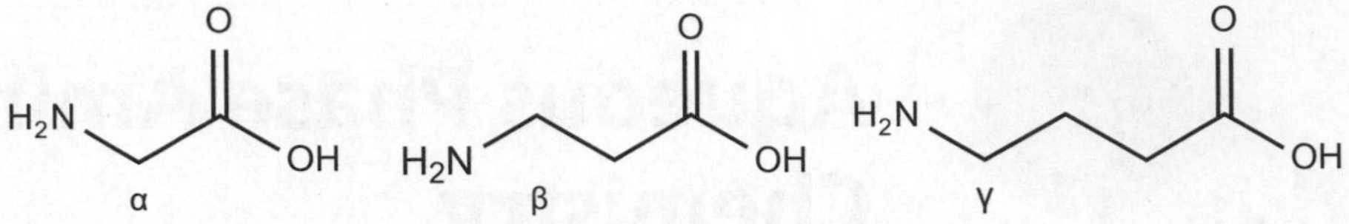

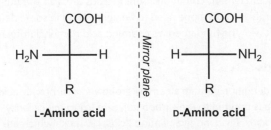

Figure 6.1 Examples of chemical structures of α-, β- and γ-amino acids.

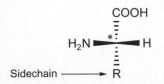

Figure 6.2 Structure of a generic α-amino acid depicting the sidechain and asymmetric carbon.

The α-carbon is an asymmetric centre

In all but one exception, the α-carbon of a proteinogenic α-amino acid is an asymmetric centre. The one exception is glycine—the simplest amino acid—whose α-carbon is bonded to two protons in addition to the amine and carboxylic acid groups, making it achiral. All other proteinogenic amino acids have a substituent on the α-carbon yielding four differently bonded groups, making the α-carbon for these amino acids asymmetric centres. These substituents on the α-carbon are referred to as sidechains, and these sidechains and configurations of any asymmetric centres determine the identity of the α-amino acids, see Figure 6.2.

Fischer projections

As we saw with sugars, we can use D/L nomenclature to assign the stereochemistry of the α-carbon. Like the sugars, we can use Fischer projections to draw the structures of the α-amino acids. The convention is to draw the carboxylic acid group at the top, the sidechain (R) at the bottom, and the asymmetric α-carbon in the middle with the horizontal lines representing the bonds to the amino group and the proton. If the amino group is on the left, it is an L-amino acid; if it is on the right, it is a D-amino acid, see Figure 6.3.

With a few exceptions (some noted in Chapter 3), biology uses L-α-amino acids almost exclusively. In comparison to sugars, the amino acids are generally less complicated in terms of their stereochemistry, but let's examine two more examples before moving on, namely, proline and threonine.

Proline is unique in that its sidechain is also bonded to its amine, forming a 5-membered ring. Nevertheless, we can still draw the Fischer projection for both D and L **isomers** as shown in Figure 6.4.

Another unique α-amino acid is threonine, because it has two asymmetric centres, the second being in the sidechain itself. Since there are two asymmetric centres,

Figure 6.3 Fischer projections of generic D- and L-amino acid enantiomers. Recall from Chapter 5 the horizontal bonds by convention are coming towards you out of the page, while the vertical bonds are going away from you behind the page.

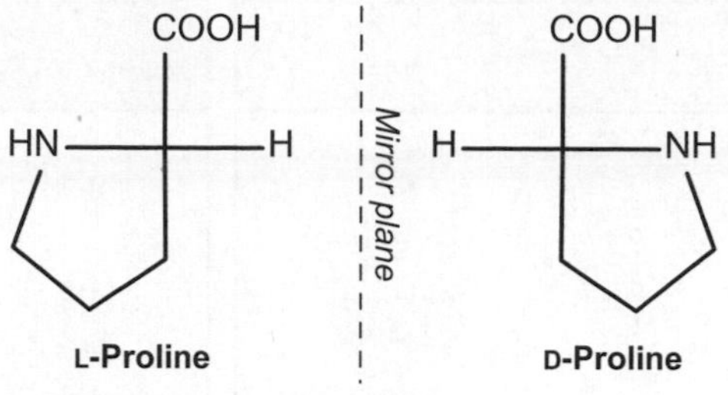

Figure 6.4 Fischer projections of D- and L-proline.

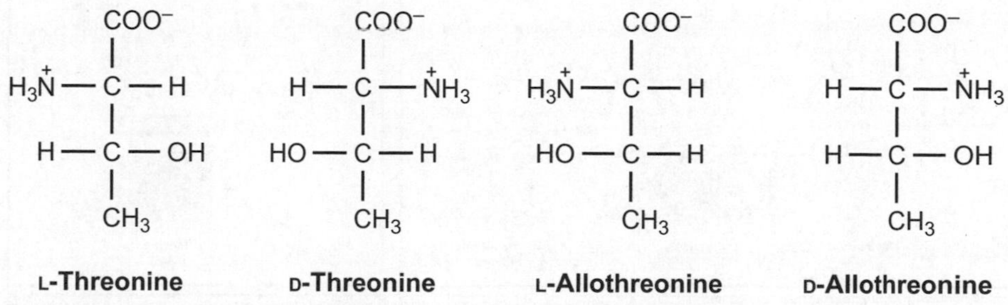

Figure 6.5 Fischer projections of D- and L-threonine and D- and L-allothreonine.

we can draw four stereoisomers as shown in Figure 6.5 (two of which are known as allothreonine when both the amino and hydroxyl groups are on the same side). Biology, however, only uses the one shown on the far left. Note we still call this an L-amino acid, because the amino group is on the left-hand side, and we ignore the fact that the OH group is on the right-hand side, even though it is the bottom-most asymmetric centre. Isoleucine is the only other proteinogenic amino acid with an asymmetric centre in its sidechain.

Sidechains and their properties

In principle, the sidechain of an amino acid can be almost any organic substituent, but biology has converged on a set of 20 α-amino acids as shown in Figure 6.6.

There are a few different ways to classify amino acids according to the polarity of their sidechains. There are the nonpolar amino acids, for example, those with hydrocarbon and aromatic sidechains; polar amino acids, like asparagine and glutamine with amide sidechains; and acidic and basic amino acids, like glutamic acid and lysine with carboxylic acid and amine sidechains, respectively. The sidechains play important roles in both the structure and enzymatic function of proteins.

All amino acids have ionizable protons, and the total charge of an amino acid depends on the pH and identity of the sidechain. The pK_a values of the carboxylic acids bonded to the α-carbons for the proteinogenic amino acids are all around 2, while the pK_a for the α-amino groups are between 9 and 10. These pK_a values mean that all amino acids that do not possess acidic or basic sidechains at pH 7 exist as zwitterions with a net-neutral charge, i.e. the amino groups are protonated, and the carboxylic acid groups are deprotonated.

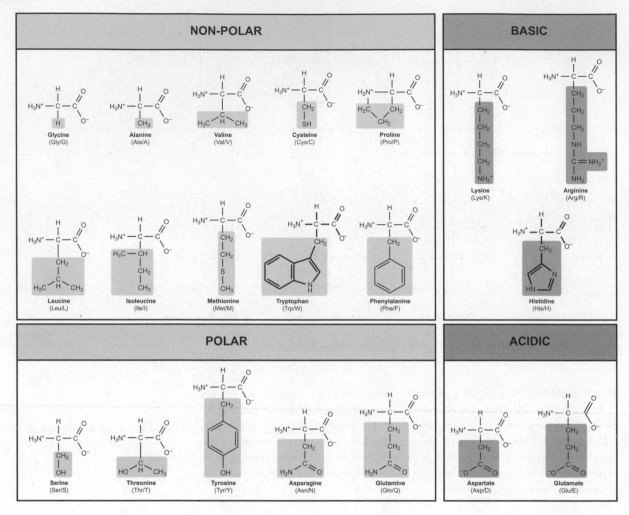

Figure 6.6 The chart of all 20 proteinogenic amino acids categorized by their sidechain properties, showing their three and one letter codes.

Peptides

Amino acids can link together in head-to-tail fashion to form peptides, which are polymers of amino acids. By convention, whenever writing the sequence of a peptide, the end terminating with an amine starts on the left and the carboxylic acid terminus on the right. The bonds that link peptides are amides, which are also referred to as **peptide bonds** in this context. These bonds form when the amine of one amino acid substitutes for the hydroxyl of the carboxylic acid of another, formally eliminating a molecule of water in the process, see Figure 6.7. The equilibrium for this reaction favours the amino acid starting material, and so to obtain appreciable quantities of peptides prebiotically, chemical or physical activation is required which will be discussed in Chapter 7.

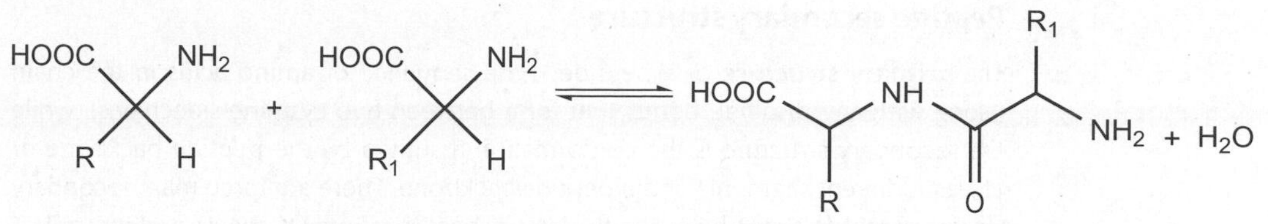

Figure 6.7 Scheme showing formation of a peptide bond from two amino acids by the elimination of water.

Properties of the peptide bond

The amides of peptides have important properties that play a major role in determining the structure of a peptide. At first glance, you may think there is free rotation about the carbon-nitrogen amide bond. Such rotation is, however, hindered because this linkage has partial double bond character. Examination of the major contributing resonance structures helps explain why, see Figure 6.8.

The resonance structure on the right shows how that double bond character arises, which imparts rigidity to the peptide backbone and has important consequences for how peptides and proteins can fold. As a consequence of the partial double bond character, these atoms all lie in the same plane, referred to as the amide plane. This means there are effectively only two degrees of rotational freedom per peptide residue (excluding possible rotations in the sidechains themselves). There is free rotation around the N-α-carbon bond and about the α-carbon-carbonyl carbon bond.

Peptide bonds can serve as both hydrogen bond donors and acceptors as shown in Figure 6.9. These hydrogen bonds afford the peptide backbone the ability to adopt a few different secondary structural motifs.

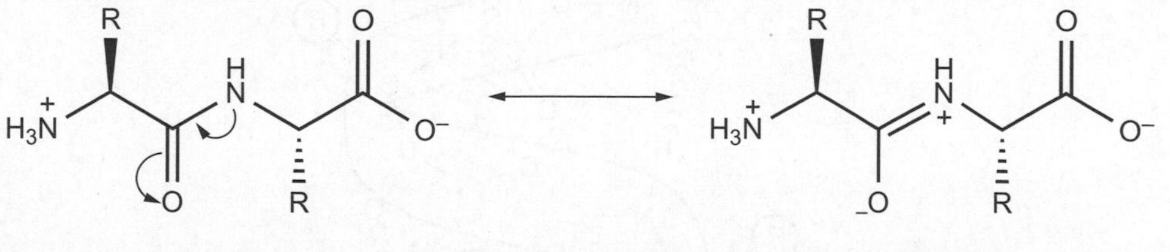

Figure 6.8 Peptide bond resonance structures illustrating the peptide bond's double-bond character.

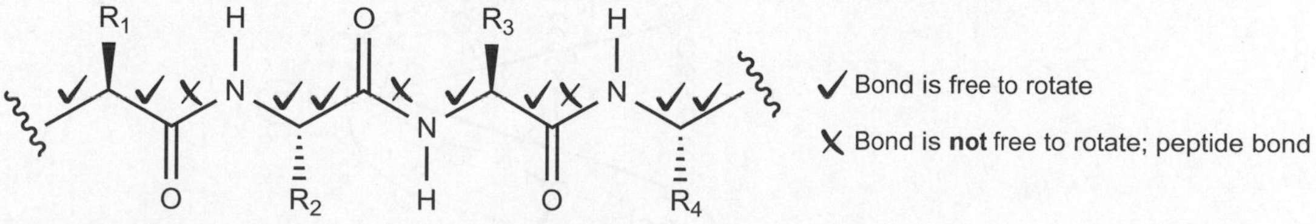

Figure 6.9 The restricted rotation of a peptide as a result of amide double-bond character.

Note: Free rotation is possible around the two single bonds to the α-carbon.

Peptide secondary structure

The **primary structure** of a peptide is the sequence of amino acids in the chain (along with any disulfide bonds that form between two cysteine sidechains), while the secondary structure is the conformation assumed by the peptide backbone or at least different segments of the peptide backbone. There are three main secondary structural motifs: the α-helix, the β-pleated sheet (or simply β-sheet), and the coil.

The α-helix, like the name suggests, is a helix (although the 'α' prefix has no direct connection to the α-carbon and is called an α-helix for historical reasons). Unlike deoxyribonucleic acid (DNA), however, it is only a single-stranded helix, which is held together by a series of intrastrand hydrogen bonds involving the backbone peptide bonds, see Figure 6.10. The sidechains of each amino acid residue point away from the helix and are found on its exterior. The carbonyl oxygen of one amide acts as a hydrogen-bond acceptor to an N–H of another amide three residues up the peptide chain. While one hydrogen bond is not especially strong, several add up in a cooperative fashion to form highly stable structures. Some amino acids have a stronger tendency than others to form α-helices. For example, polyalanine readily forms an α-helix, but polyaspartate or polylysine do not due to the high intramolecular charge-charge repulsion that would result.

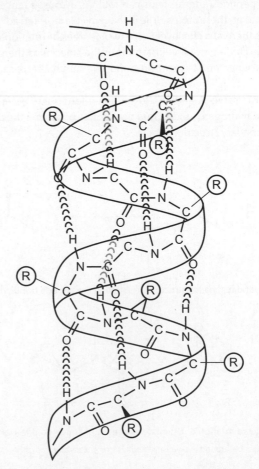

Figure 6.10 Structure of the peptide α-helix held together with hydrogen bonding.

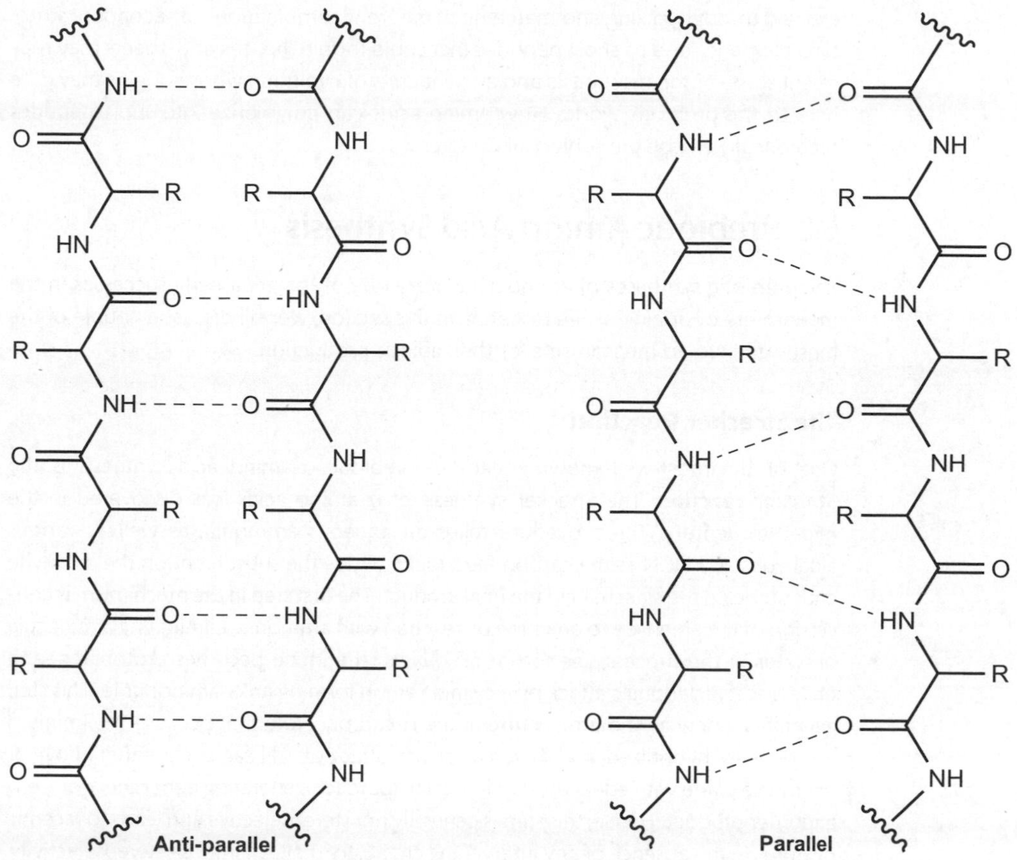

Figure 6.11 Structure of peptide β-sheets held together with hydrogen bonding.

The β-pleated sheet can be visualized as a sort of two-dimensional (2D) sheet of pleated paper forming a zig-zag pattern. Like the α-helix, β-pleated sheets are held together by hydrogen bonds between backbone amides, but unlike α-helices, these hydrogen bonds are *inter*strand rather than *intra*strand. The strands can be arranged either parallel to each other, running in the same direction, or antiparallel, running in opposite directions, see Figure 6.11. The sidechains of β-sheets are distributed on either side of the 2D sheet. β-Sheets in proteins are usually found with multiple strands, but even small peptides are known to form two-stranded antiparallel β-sheets thanks to what's called a β-turn. β-Turns allow a peptide strand to make a sharp bend or 'U-turn' enabling it to reverse direction. Proline residues often help facilitate β-turns.

Beyond secondary structure, there is also tertiary and **quaternary structure**, which describe how proteins fold into complex three-dimensional (3D) shapes based on secondary structural units, and then how multiple proteins bind with each other in an intramolecular fashion, respectively. We will end our discussion with secondary structure, however, as tertiary and quaternary structures are complicated topics in the domain of protein chemistry and beyond the scope of this discussion in prebiotic chemistry. It is hypothesized, however, that short peptides may have arisen spontaneously on early Earth through abiotic processes, and some of these peptides could have displayed basic secondary structural motifs. Proteins are thought to have evolved through mixing and matching of different combinations of secondary structural elements, and so short peptides that could form α-helices or β-sheets may represent some of the most basic and ancient units of evolution whose origins may date back to the prebiotic world. How amino acids can polymerize into short peptides prebiotically will be the subject of Chapter 7.

6.2 Prebiotic Amino Acid Synthesis

The prebiotic syntheses of amino acids were one of the great early successes in the modern era of origins-of-life research. In this section, we will discuss a couple of the most well-studied mechanisms for their abiotic production.

The Strecker Reaction

One of the most-well-known means of prebiotic α-amino acid synthesis is the **Strecker reaction**. The Strecker synthesis of α-amino acids was discovered in the mid-19th century. The procedure relies on aqueous ammonia, as well as various aldehydes and HCN as the carbon feedstocks, where the substituent on the aldehyde ends up being the sidechain in the final product. The first step in the mechanism is conversion of the aldehyde to an imine by reaction with ammonia, eliminating a molecule of water in the process, see Figure 6.12. Next, the imine becomes protonated, and undergoes nucleophilic attack by a cyanide anion forming an α-aminonitrile. This step generates a new asymmetric centre at the α-carbon. Note that because the iminium carbon is sp^2 hybridized, and therefore planar, attack of ⁻CN can occur either above or below the plane with equal probability, each approach generating either the D or L enantiomer product. Hence, this step is generally not stereoselective and leads to racemic mixtures in the absence of any intervening chemistry that can impose stereoselectivity. This step may also remind you of the Kiliani–Fischer synthesis of sugars discussed in Chapter 5 and reveals a mechanism by which sugars and amino acids can arise from a common mixture.

The final step involves either an acid-catalysed or base-promoted hydrolysis of the nitrile to the carboxylic acid, generating a molecule of NH_3 in the process, as described in the following mechanisms. This hydrolysis is typically achieved by significantly raising or lowering the pH of the reaction mixture in a separate follow-up step. In the acid-catalysed reaction, the first step is protonation of the nitrile nitrogen, activating its carbon for nucleophilic attack in the next step by water, see Figure 6.13A. After proton transfer and tautomerization, an amide group is formed, resulting in the relatively stable α-aminoamide intermediate. Another protonation, this time at the carbonyl oxygen, is required in order to prime the carbonyl carbon of the amide

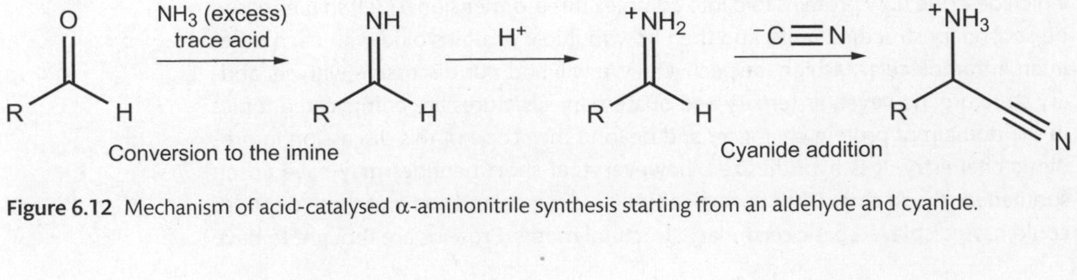

Figure 6.12 Mechanism of acid-catalysed α-aminonitrile synthesis starting from an aldehyde and cyanide.

for nucleophilic attack by another molecule of water. After addition of water, two successive proton transfer steps lead to an ammonium diol tetrahedral intermediate, which collapses back down to reform the carbonyl π-bond, eliminating NH_3 at the same time. A final proton transfer yields the carboxylic acid. In the base-promoted mechanism, the initial step is direct attack on the nitrile carbon by ^-OH, which after proton transfer and tautomerization yields the same α-aminoamide intermediate as before, see Figure 6.13B. The amide carbonyl carbon is attacked again by another ^-OH forming a negatively charged oxide tetrahedral intermediate, which collapses back down eliminating $^-NH_2$ (which is concurrently protonated to NH_3), while yielding the carboxylic acid. In the final step, the acid is deprotonated to the carboxylate.

The Strecker reaction lends itself well to prebiotic synthesis of amino acids for a couple of reasons: (*i*) HCN and aldehydes were likely abundant on early Earth, and (*ii*) the first few steps leading to the α-aminonitrile intermediates are all reversible

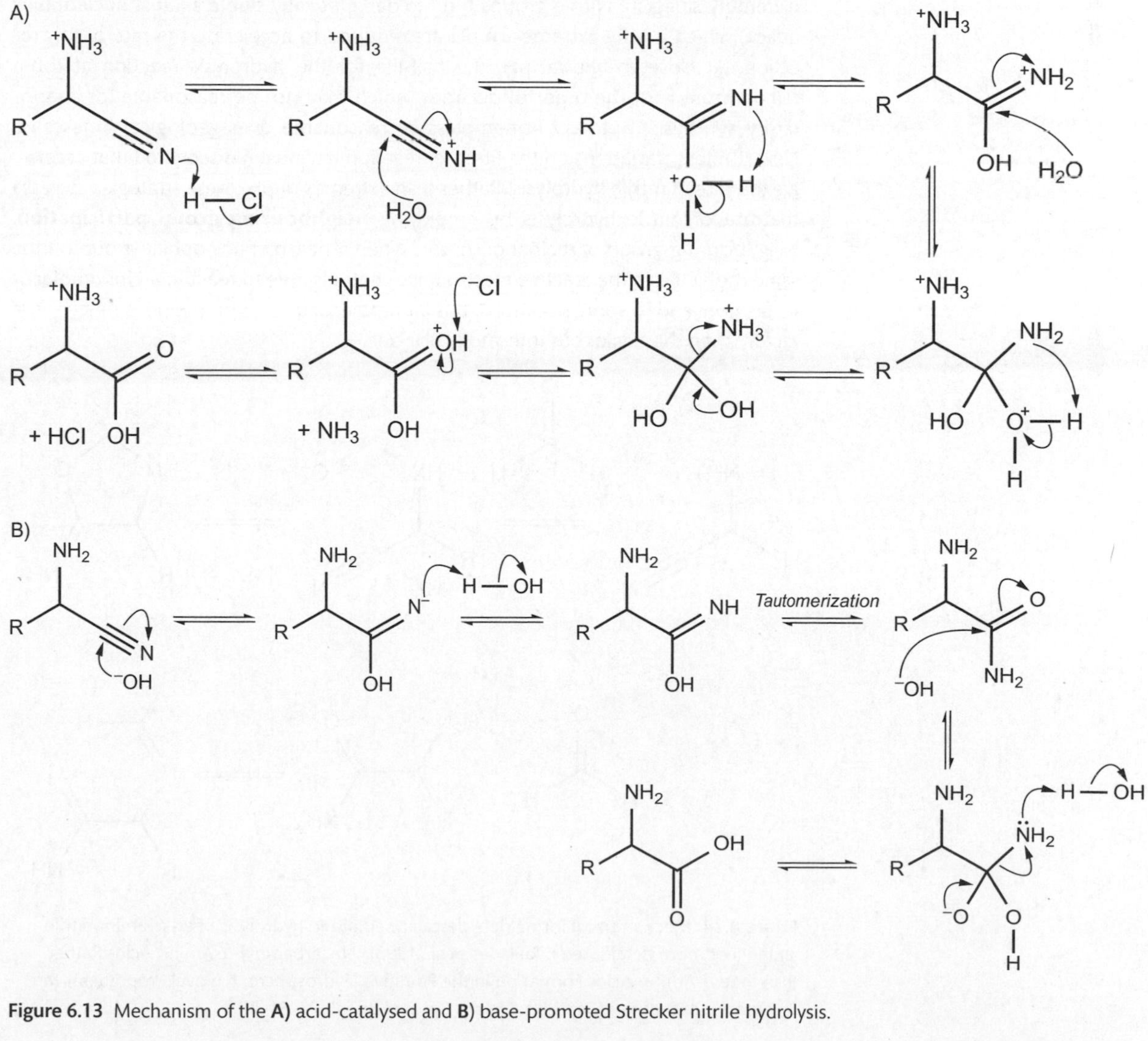

Figure 6.13 Mechanism of the **A)** acid-catalysed and **B)** base-promoted Strecker nitrile hydrolysis.

and can take place in the same reaction mixture at slightly basic pH. Meanwhile, the cyanide anion can also directly attack the aldehyde instead of the iminium, leading to the cyanohydrin derivative instead (as we already saw in the Kiliani–Fischer synthesis of sugars), and the proportion of cyanohydrin to α-aminonitrile obtained depends in large part on the NH_3 concentration and pH. α-Aminonitriles are generally favoured at more basic pH due to the requirement of needing free NH_3 (as opposed to the non-nucleophilic NH_4^+) to form the imine. There are situations where α-aminonitriles can form even in the absence of any initial aqueous ammonia; if the iminium is derived by the reduction of a nitrile, like we saw is possible in Chapter 5, then α-aminonitriles also can form.

Variations of the Strecker reaction

The rate of nitrile hydrolysis at room temperature and near neutral pH is, however, extremely sluggish. Nitrile groups tend to be kinetically stable against nucleophilic attack, which is why extremes of pH are required to accelerate the rate of this reaction. At close-to-neutral pH, the half-life for the hydrolysis reaction at room temperature is of the order of decades, which may not be reasonable for a laboratory synthesis, but may nonetheless be reasonable over geological timescales. Nevertheless, variations on the Strecker reaction have been discovered that accelerate the rate of nitrile hydrolysis. Rather than extremes of pH, these strategies increase the rate of nitrile hydrolysis by employing **neighbouring group participation**. Neighbouring group participation occurs when a nearby nucleophilic group on the same molecule as the reactive electrophilic centre is able to react in an *intramolecular* fashion, and in some situations, this intramolecular reaction happens much more rapidly than the analogous intermolecular one.

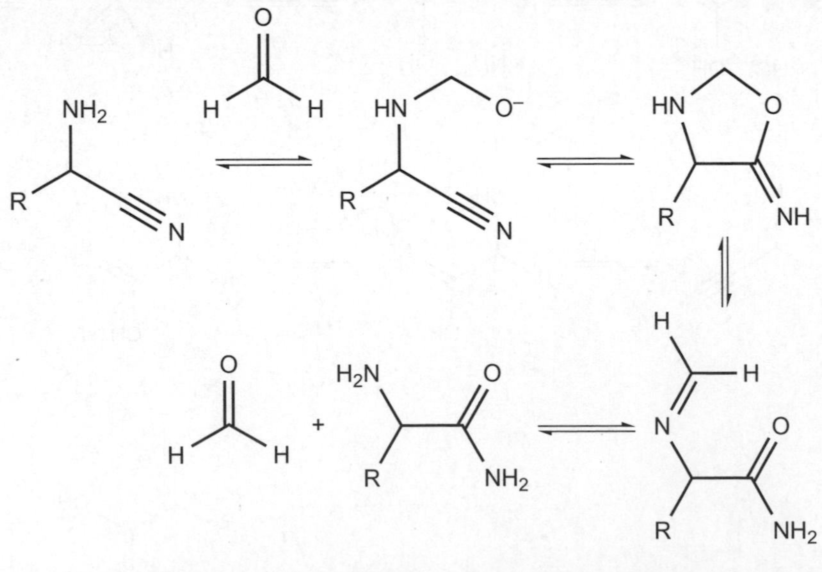

Figure 6.14 Mechanism of formaldehyde-catalysed nitrile hydrolysis of an α-aminonitrile amine. For more details, see J. Taillades et al. (1998). 'N-Carbamoyl-α-Amino Acids Rather than Free α-Amino Acids Formation in the Primitive Hydrosphere: A Novel Proposal for the Emergence of Prebiotic Peptides'. *Origins Life Evol Biosph 28*: 61–77.

For example, formaldehyde can catalyse the first steps of α-aminonitrile hydrolysis leading to the amide intermediate in a mechanism that employs a neighbouring group participation mechanism, see Figure 6.14. In the first step, the amine of the α-aminonitrile adds to the carbonyl carbon of formaldehyde, forming a hemiaminal intermediate. In slightly basic conditions, the alcohol group deprotonates, forming a strong nucleophile, and by virtue of being in close proximity to the nitrile, rapidly attacks forming a five-membered ring. Note, five-membered rings generally have low ring strain and thus form favourably, another feature that facilitates this cyclization. This ring then opens up via hydrolysis to form the α-aminoamide as an imine derivative. In the next steps, the imine undergoes hydrolysis, restoring the formaldehyde catalyst. This sort of neighbouring group participation has also been referred to as *temporary intramolecularity*, because of this catalytic nature of the formaldehyde. In comparison to nitriles, hydrolysis of amides to carboxylic acids under mildly basic pH is relatively rapid.

In another example, CO_2 instead of formaldehyde can carry out a similar function. The mechanism begins with an attack of the α-aminonitrile amine on the carbon of the CO_2, forming a carbamate intermediate, see Figure 6.15. Like before, neighbouring group participation and the fact that five-membered rings are favoured facilitates the following cyclization reaction, resulting from attack of the carbamate oxide on the nitrile carbon. This ring then opens back up forming the amide as before, but with an isocyano group. This isocyano group is highly reactive as a carbon-centred electrophile, and even the amide nitrogen, which is normally considered a weak nucleophile can attack, forming another five-membered ring, known as a hydantoin, extremely rapidly. These rings are relatively stable but will undergo hydrolysis slowly to form N-carbamoyl-α-amino acid derivatives, and eventually α-amino acids. This type of mechanism that uses CO_2 and α-aminonitriles to furnish hydantoin intermediates is known generally as the Bucherer–Bergs synthesis.

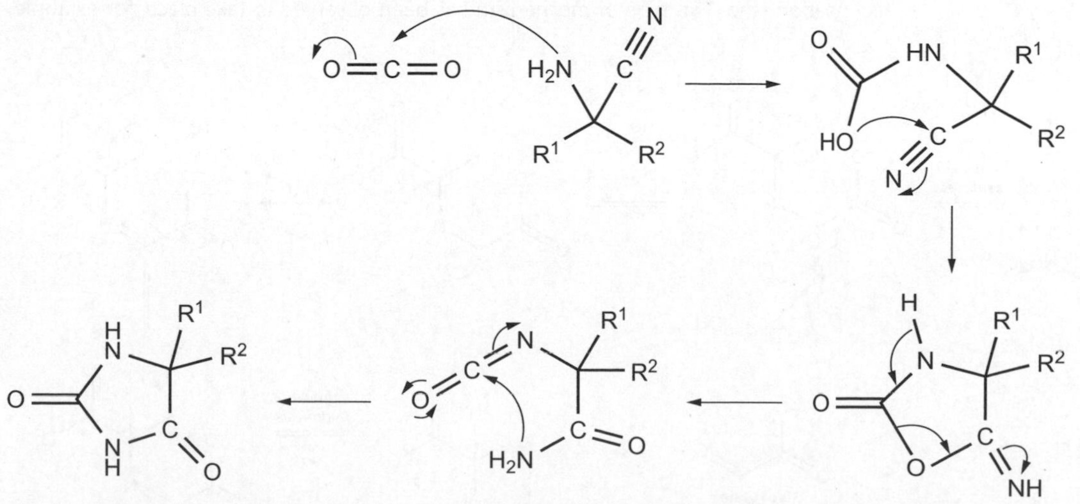

Figure 6.15 Mechanism of CO_2-catalysed nitrile hydrolysis of an α-aminonitrile amine. For more details, see J. Taillades et al. (1998). 'N-Carbamoyl-α-Amino Acids Rather than Free α-Amino Acids Formation in the Primitive Hydrosphere: A Novel Proposal for the Emergence of Prebiotic Peptides'. *Origins Life Evol Biosph* 28: 61–77.

Transaminations of α-oxoacids

Although the Strecker reaction and its variations are straightforward ways to make α-amino acids from simple feedstocks thought to be readily available on early Earth, this synthetic pathway stands in stark contrast to the metabolic pathways employed by biology discussed in Chapter 4. In particular, biosynthesis at no point involves cyanide or hydrolysis of nitriles. Instead, for all its amino acids, biology employs a protocol of transamination to install the α-amino group, using an α-oxoacid as a substrate. There are in fact potentially prebiotic mechanisms which can produce amino acids in this way. These sorts of transamination reactions can still occur even without enzymes, albeit at much slower rates. Let's examine the case of the transamination reaction of glycine and pyruvate which furnishes alanine to see how the mechanism works.

Like many reactions we have seen, the first step is addition of the glycine α-amine to the ketone carbonyl of pyruvate, which after a couple more steps, forms the imine intermediate (Figure 6.16). In the next step, the glycine-derived α-carbon is deprotonated, and the resulting negative charge can be delocalized to the pyruvate-derived α-carbon as shown in the resonance structures in Figure 6.16. This carbon becomes protonated, swapping the carbon involved in the imine bond. Note, this reaction is a type of tautomerization, and can also be referred to as a 1,3-proton transfer. In the final step, the imine hydrolyses, yielding alanine and glyoxylate.

Metal cations like Co^{2+}, Ni^{2+}, Cu^{2+}, and V^{5+} can also help catalyse this transamination reaction. The imine intermediates that form are capable of coordinating these different metals as shown in Figure 6.17. For V^{5+} metal complexes, the acidity of the α-proton becomes markedly increased, which increases the rate of tautomerization that is the rate-limiting step. For Cu^{2+} and Co^{2+}, these metal complexes stabilize the reactive imine intermediates, species of which are generally not thermodynamically favourable in water but complexing with these metals effectively increases their availability. Ni^{2+} complexes work by a combination of these two mechanisms, see Figure 6.17.

Let's examine one more mechanism for transamination that involves a decarboxylation step. This type of mechanism has been observed to take place, for example,

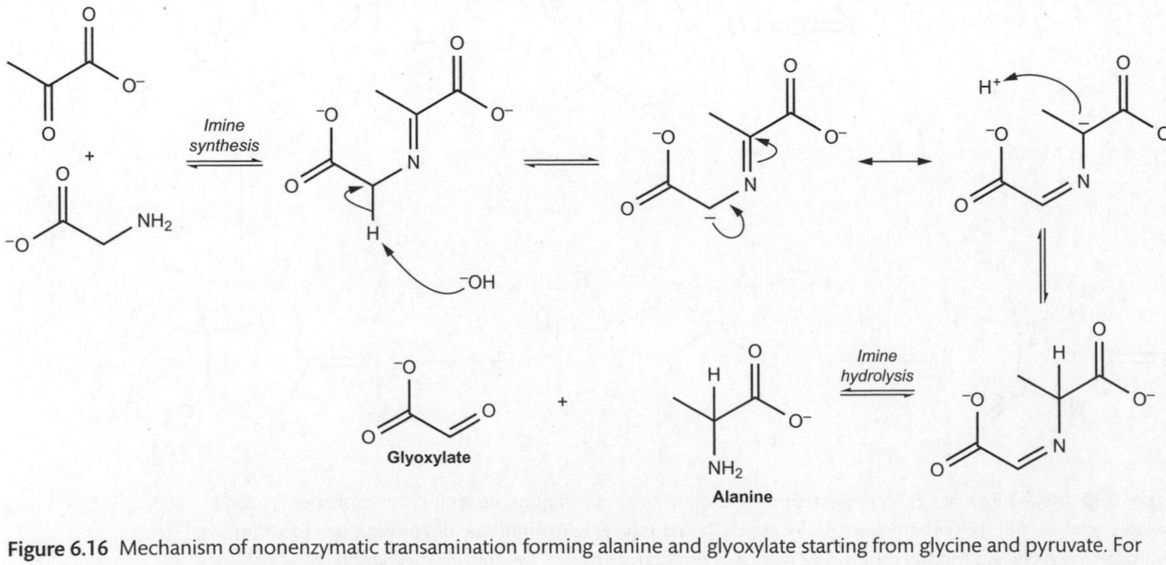

Figure 6.16 Mechanism of nonenzymatic transamination forming alanine and glyoxylate starting from glycine and pyruvate. For more details, see M. Conley et al. (2017). 'Reaction of Glycine with Glyoxylate: Competing Transaminations, Aldol Reactions, and Decarboxylations'. *J Phys Org Chem* 30: e3709.

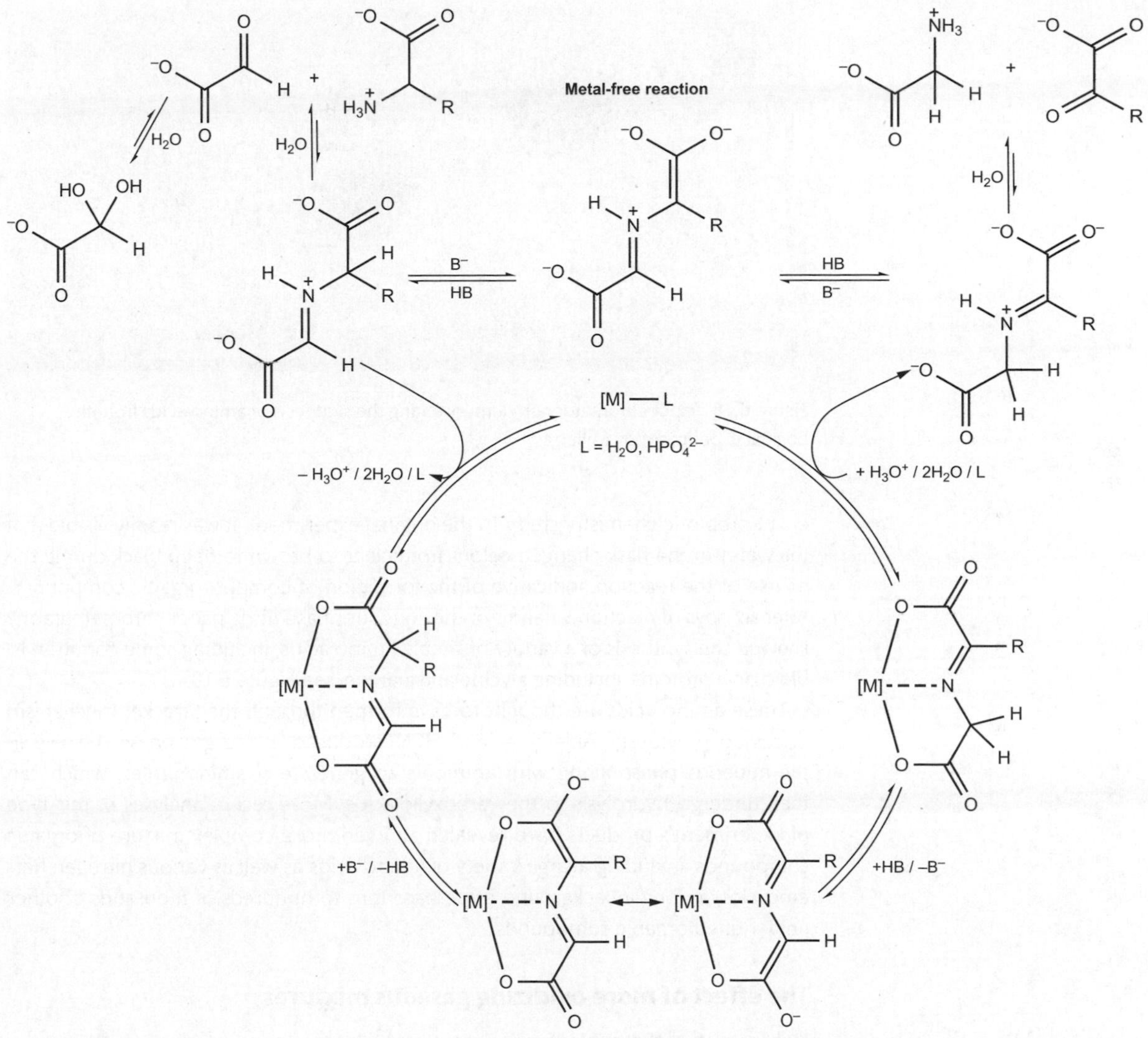

Figure 6.17 Mechanism of metal-catalysed transamination involving glyoxylate and an amino acid. For more details, see R. J. Mayer et al. (2021). 'Mechanistic Insight into Metal Ion-Catalyzed Transamination'. *J Am Chem Soc 143*: 19099–19111.

when phenylalanine reacts with pyruvate. As before, the first step is formation of an imine intermediate. Instead of tautomerization, however, decarboxylation of the phenylalanine-derived carboxylic acid takes place, which results in another imine whose double bond has migrated to the neighbouring carbon. Hydrolysis of this imine intermediate results in the amino acid alanine along with phenylacetaldehyde.

6.3 Miller–Urey-Type Experiments

As discussed in Chapter 2, the Miller–Urey experiment, which demonstrated that mixtures of reducing gases when subjected to electrical discharge can produce amino acids, was one of the remarkable discoveries that helped initiate the modern

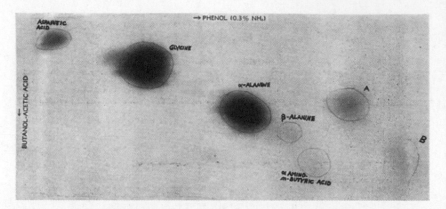

Figure 6.18 Paper chromatogram demonstrating the synthesis of amino acids from the original experiment by Miller.

era of prebiotic chemistry study. In the original experiment, it was readily visible that the water in the flask changed colour from clear to brown to deep black during the course of the reaction, indicative of the formation of complex organic compounds. After six days of reaction, analysis of the aqueous phase using paper chromatography showed the synthesis of a variety of simple amino acids, including some common to biological proteins, including glycine and alanine, see Figure 6.18.

These amino acids are thought to have formed through the Strecker mechanism discussed previously. Aldehydes and HCN produced in the gas phase dissolve in the aqueous phase along with ammonia to generate α-aminonitriles, which can then undergo hydrolysis to the carboxylic acids. More recent analyses of this type of experiment's products have revealed an even more complex mixture of organic compounds, including a large variety of amino acids as well as various nitrogen heterocycles, among a background of perhaps tens to hundreds of thousands of other unidentified organic compounds.

The effect of more oxidizing gaseous mixtures

Not long after the original experiment by Miller was conducted, some doubt was cast as to whether the primitive atmosphere was especially reducing. This experiment has since been repeated with various gas mixtures ranging from highly reducing to neutral (e.g. CO_2, N_2, H_2O) to oxidizing. In general, it has been found that the more reducing the gas mix, the more abundantly and efficiently organics are produced, while extremely oxidized gas mixes do not typically produce abundant organics, and specifically, amino acids in great abundance or high diversity.

Nevertheless, since Miller's pioneering experiment, the general phenomenon of organic synthesis occurring in natural planetary environments has been confirmed. For example, the atmosphere of Saturn's moon Titan consists of a dense organic haze essentially produced from gas-phase reactions of methane and N_2, and the outer planets and various extraterrestrial materials including comets, asteroids, and meteorites have been found to host organics similar to those observed in such experiments, including amino acids, which brings us to the final topic of this chapter, amino acids in meteorites.

6.4 Amino Acids in Meteorites

As discussed in Chapter 2, extraterrestrial (ET) input from comets, meteorites, and interstellar dust particles (IDPs) provides a mechanism by which organic matter could have been delivered to the early Earth, including a variety of proteinogenic and non-biological amino acids, among others. Large numbers of meteorites have now been recovered and their chemistries studied, and it has been discovered that in many cases, there is often a measurable enantiomeric excess of L over D amino acids for certain amino acid isomers. It has been postulated that meteorites may have provided the original enantiomeric excesses that were ultimately amplified by biology and offer an explanation to the origin of homochirality. In this next section, we will briefly describe models for amino acid synthesis in space on meteorite parent bodies, the Murchison meteorite, and progress in sample return missions.

Amino acid synthesis in meteorite parent bodies

Meteorites are the rocky remnants of asteroids or comets that have landed on Earth, and as such, they can provide insight into the organic chemistry of the early Solar System. Meteorites have a long history that dates back to the original molecular cloud from which the Solar System formed. During the accretion of the rockier types of small bodies in the early Solar System (e.g. asteroids), radiogenic nuclides generated in the precursor supernova (most importantly ^{26}Al) released significant amounts of heat which was inhibited from convectively escaping as the bodies grew in size. This trapped heat led to melting and differentiation of the accreted materials from the inside out, with denser materials sinking to small body interiors, as well as extensive melting and circulation of water and its contents within the parent bodies. During this phase, simple reactants including HCHO, NH_3, and HCN, which we will recall were similarly produced in the Miller–Urey experiment, were brought into low-temperature aqueous conditions that allowed them to react similarly. Indeed, significant similarities between the organic products of Miller–Urey-type reactions and the contents of carbon-rich meteorites have been noted. Thus, even in the absence of a significantly reduced early terrestrial atmosphere, similar organic products could have been brought to Earth by ET materials.

The Murchison meteorite

A particular class of meteorites, the carbonaceous chondrites, have largely dominated perceptions of meteoritic contributions to the early Earth organic inventory, partly because such materials are macroscopic and have been supplied in quantities sufficiently large to enable extensive characterization. An excellent example of this is the Murchison meteorite, which fell over Murchison, Australia in late 1969, see Figure 6.19.

The Murchison meteorite marked a major shift in our understanding of the distribution of organic materials in the Solar System. Until this meteorite's fall, it was expected that organic materials might be widely distributed in the Solar System, yet analysis of lunar materials returned by the Apollo missions revealed the Moon to be almost completely devoid of organic material. This was disappointing for scientists engaged in exobiological research at the time, but Murchison provided the needed counterpoint.

Figure 6.19 Photograph of a fragment of the Murchison meteorite.

Meteorites can be classified as 'falls' (those observed to fall) and 'finds' (those found having fallen unobserved). Murchison was observed to fall, and ~70 kg of Murchison was collected shortly after its impact, which greatly lessened the likelihood of contamination compared to most 'finds' which may have spent unknown time periods on Earth's surface. Murchison is now known to represent an especially carbon-rich type of meteorite and yielded extensive new information about the nature of ET organic materials. Murchison is among the most organic-rich types of meteorites, with the diversity of organics in meteorites believed to be related to the maturational history of their parent bodies, in terms of the length and temperature at which they were radiogenically heated. Moderate maturation is associated with extensive molecular diversification, while more extensive maturation is associated with destruction of organic material. It is worth noting that since the proliferation of real-time sky monitoring, both deliberate and inadvertent in the form of dashcams and so on, the frequency of observation of infalling ET bodies has increased significantly. Indeed, such events are recorded almost daily via local news.

Sample return missions

Scientists have now studied a great number of comets and asteroids both spectroscopically and via *in situ* and sample return measurements, and their organic content has been more deeply characterized. In addition to the meteorites collected

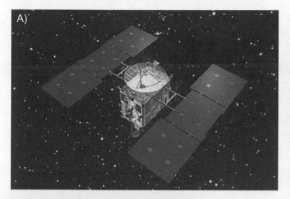
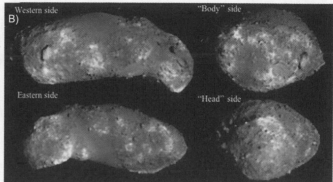
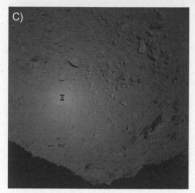

Figure 6.20 Image of **A)** the Hayabusa probe and **B)** and **C)** the Itokawa asteroid.

Note: The second Hayabusa mission revealed an organic suite highly reminiscent of previous carbonaceous chondrite (CC) analyses.

on the ground and observed to fall, there is increasing evidence for the input of small comet-like bodies via satellite observations of Earth. It seems likely the input of small comet-like bodies must be relatively common over geological timescales, though such events leave little recoverable evidence.

The 21st century has witnessed numerous sample return missions from asteroids. Samples returned so far from Japan's first Hayabusa mission (Figure 6.20) revealed extensive graphite, but little organic carbonaceous material, which can perhaps be ascribed to the sampling depth; the surface materials (from perhaps only the top few cm of the asteroid) collected by this mission have been exposed to extensive space weathering, and thus are extremely altered. More recent sampling has revealed organics more typical of the Stardust mission findings, and scientists are thus slowly putting together a picture of the continuum of organics produced in ET settings.

6.5 Summary

- Amino acids contain an amino group and a carboxylic acid. Proteinogenic amino acids are all α-amino acids.

- With the exception of glycine, the α-carbon of proteinogenic amino acids are asymmetric centres. Biology uses L-α-amino acids almost exclusively.

- Important prebiotic amino acid syntheses include the Strecker reaction, variations thereof, and transamination, which can be catalysed by transition metals.
- Amino acids have been detected in meteorites, sometimes with an L-enantiomeric excess.
- Meteorites may have provided the original enantiomeric excesses that were ultimately amplified by biology, offering an explanation to the origin of biological chirality.

6.6 Exercises

1. Do you expect the Strecker synthesis of amino acids to produce more of the D or L enantiomers? Draw a reaction mechanism to justify your answer.

2. Part of the mechanism for the formaldehyde-catalysed nitrile hydrolysis for α-aminonitriles involves the formation of a five-membered ring as an intermediate. Why would the formation of a five-membered ring be more favoured than an attack by ⁻OH?

3. Do you think the transamination mechanism for prebiotic amino acid synthesis is stereoselective? Explain.

4. Do you think β- and γ-amino acids can be synthesized by the Strecker mechanism? Why or why not?

5. Which amino acid synthesis mechanism discussed in this chapter do you think is the most prebiotically plausible? Explain your reasoning and provide a comparison to extant biochemical pathways.

6.7 Suggested Reading

1. Moran Frenkel-Pinter, Mousumi Samanta, Gonen Ashkenasy, and Luke J. Leman. (2020). 'Prebiotic Peptides: Molecular Hubs in the Origin of Life'. *Chem Rev 120*: 4707–4765.

2. Jamie E. Elsila, José C. Aponte, Donna G. Blackmond, Aaron S. Burton, Jason P. Dworkin, and Daniel P. Glavin. (2016). 'Meteoritic Amino Acids: Diversity in Compositions Reflects Parent Body Histories'. *ACS Cent Sci 2*: 370–379.

3. Norio Kitadai and Shigenori Maruyama. (2018). 'Origins of Building Blocks of Life: A Review'. *Geosci Front 9*: 1117–1153.

4. Dimas A. M. Zaia, Cássia Thaïs B. V. Zaia, and Henrique De Santana. (2008). 'Which Amino Acids Should Be Used in Prebiotic Chemistry Studies?'. *Orig Life Evol Biosph 38*: 469–488.

5. Meierhenrich, U. J., Muñoz Caro, G. M., Schutte, W. A., Barbier, B., Arcones Segovia, A., Rosenbauer, H., Thiemann, W. H.-P., and Brack, A. (2002). 'The Prebiotic Synthesis of Amino Acids—Interstellar vs. Atmospheric Mechanisms'. *Proc. Second Eur. Workshop Exo/Astrobiol*. ESA SP-518.

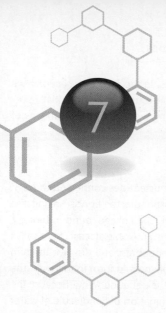

7 Nonenzymatic Polymerization of Ribonucleic Acids and Peptides

The polymerization of amino acids and ribonucleotides into long chains is what gives them their characteristic enzymatic and genetic properties. Nonenzymatic polymerization is thought to be a crucial step towards the origins of life. This chapter will focus on how amino acids and ribonucleotides can be linked together into oligomers and polymers via abiological condensation reactions. The two main strategies to achieve condensation of amino acids and ribonucleotide monomers into oligomers/polymers are (*i*) cycles of wetting and drying and (*ii*) chemical activation with condensation agents. Examples of both strategies are discussed. The use of nonenzymatic template-directed synthesis as a means of ribonucleic acid (RNA) and peptide copying is presented.

7.1 Nonenzymatic Polymerization Is Thought to be a Prerequisite for Life's Emergence

Polymers are essential to life, in particular, those expressed in specific sequences of their component monomeric building blocks. For deoxyribonucleic acid (DNA), the particular sequence of its A, T, G, and C component nucleotides is the mechanism by which the cell is able to store information. In the case of proteins, it is the primary sequence of amino acid residues which ultimately determines their enzymatic activity and function. Ribonucleic acid (RNA) holds a special place in origins-of-life research, because its sequence provides *both* a potential mechanism for storing information *as well as* carrying out catalysis and other functions. In extant biology, however, the synthesis of these biopolymers including RNA requires a suite of sophisticated (highly evolved) proteins as well as the ribosome, which *a priori* would not have existed during the early stages of prebiotic chemistry.

If the abiotic production of RNA (or some other structurally related polymer) and short peptides was a prerequisite for life's emergence, then nonenzymatic mechanisms for the polymerization of ribonucleotides and amino acids must have existed. In particular for RNA, if it were also the original information storage molecule to initially emerge geochemically, it would have also needed a nonenzymatic mechanism for its replication, i.e. a synthetic protocol by which any arbitrary sequence can be copied chemically without the aid of enzymes. In the following sections, we will discuss the fundamentals of nonenzymatic polymerization, their potentially pre-biotic protocols, and the mechanism of template-directed synthesis as a means of sequence copying.

The thermodynamics of polymerization

As discussed in Chapter 5, RNA is a polymer of ribonucleotides connected through the sugar-phosphate backbone by a series of phosphodiester bonds linking the 3'- and 5'-positions of the ribose component. (This is the linkage found in biology, but connections through the 2'- and 5'-positions are also possible at least chemically.) However, if you were to take some ribonucleotide monomers and dissolve them in water, you would never get significant amounts of polymers to form in this way—the equilibrium would vastly favour the left-hand side of the equation. Why? Because the formation of a phosphodiester bond requires the elimination of a molecule of water as shown in Figure 7.1, which is a thermodynamically unfavourable reaction charac-terized by a positive increase in free energy.

The formation of peptides in water from amino acid monomers is similarly ther-modynamically unfavourable. As for RNA, the formation of the peptide (amide) bond requires the elimination of a molecule of water, hence, significant quantities of peptides cannot be obtained through this simple mechanism. As a reference, the

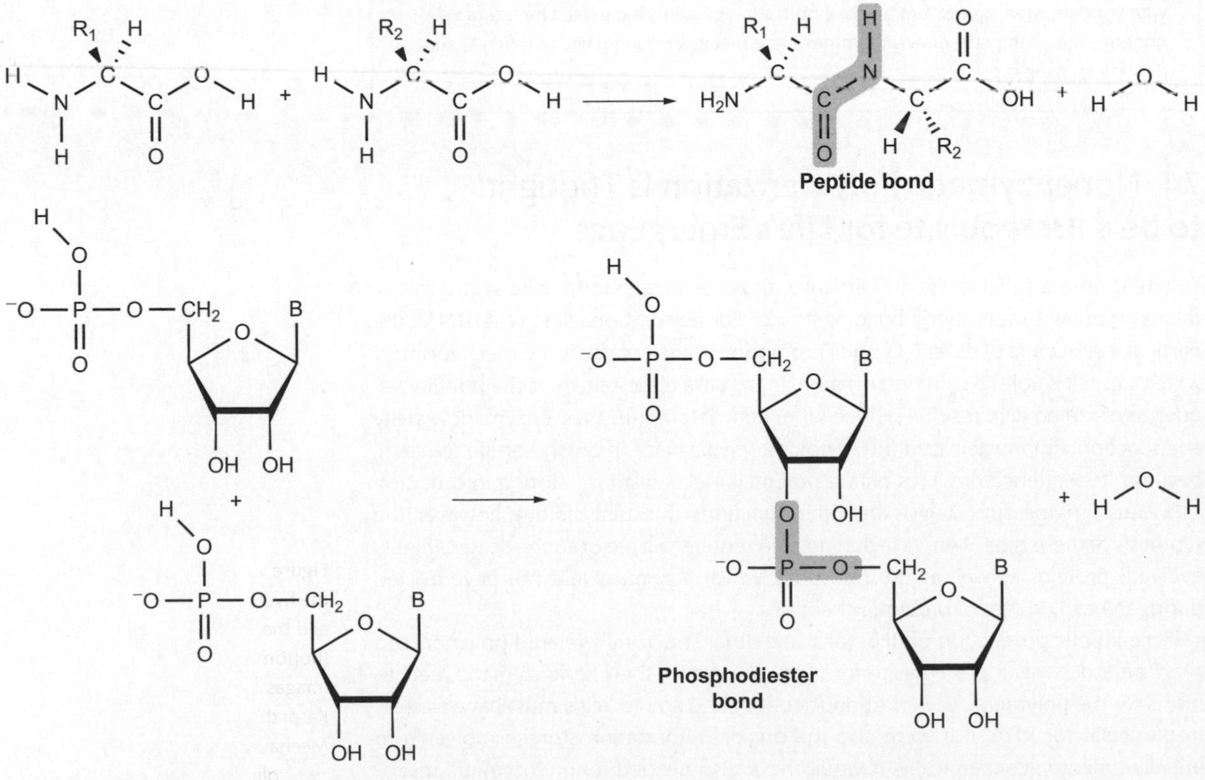

Figure 7.1 Scheme depicting general condensation reactions for making peptides (top) and RNA (bottom).

standard change in free energy at pH 7 and 25 °C for the reaction 2 Gly → GlyGly + H_2O is 3.5 kcal/mol and that for adenosine + HPO_4^{2-} → AMP + H_2O is 2.7 kcal/mol. Not only are these reactions characterized by an increase in free energy, but since water is also the solvent, it exists in a huge excess, according to Le Chatelier's principle, mass action will drive the reaction towards the reactants.

Recall from Chapter 5 that the phosphodiester bonds of RNA are easily cleaved, and in fact, this cleavage reaction is spontaneous thanks in large degree to the thermodynamic consequences of water. Although more kinetically stable than RNA, peptides too will eventually hydrolyse back to their monomeric amino acid building blocks. The fact that water favours the decomposition of RNA and peptides back into their monomer components, as well as other hydrolysis reactions in prebiotic chemistry, is sometimes referred to as the *water problem*. There is every indication, however, that water is necessary for life, but at the same time, water is constantly working against polymerization. So, what can be done to achieve polymerization on a planet mostly covered by water?

Polymerization by dry down

The most obvious way to drive a condensation polymerization reaction forward would be to remove water from the equation. This removal can be done in practice by allowing water to evaporate, thereby subtracting it from the equilibrium. On early Earth, this may have been achieved through evaporation of the aqueous mixtures containing amino acids or ribonucleotides. For example, imagine small tidal pools undergoing cycles of drying and wetting, or a lake whose shoreline recedes by evaporation, but is replenished at regular intervals by rainfall. Although such mechanisms would not have been possible in a prebiotic world completely covered by water, an early Earth with any sort of emergent landmasses would have inevitably possessed aqueous reservoirs, large or small, undergoing (semi)periodic or episodic evaporation and rewetting. Amino acids and ribonucleotides can both be polymerized by this mechanism.

> ## A note on equilibrium
>
> To say that peptide bonds cannot form in water by direct condensation is not entirely true. The equilibrium constant for peptide bond formation tends to increase at higher temperatures and pressures, and so by starting from large concentrations of amino acid monomers (~0.1 M or more) under these conditions, a small fraction can be converted to short peptides. The yields of any individual peptide oligomer, i.e. 2mer, 3mer, 4mer, etc., however, are low, ~1 mM at most under optimal circumstances. Experiments have also been conducted to see if RNA oligomers can be produced in aqueous solutions from ribonucleotide monomers under elevated pressures and temperatures, and although small oligomers can be made in this way, the yields are extremely low. Without any sort of chemical assistance or additives, making long RNA or peptide polymers in water is indeed a difficult and unlikely prospect.

Early examples of amino acid polymerization carried out in the 1950s in fact didn't involve water at all, but anhydrous melts of the amino acid monomers themselves. In these experiments, amino acids were combined as solids and heated without water upwards of 170 °C for a few hours. Analysis of the products showed the formation of

Glycine

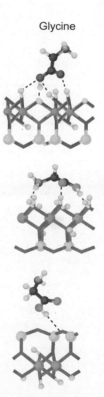

Figure 7.2 Absorption of glycine onto the hydroxyl surface (top and middle) and the siloxane surface of kaolinite (bottom) in different configurations. Images reproduced from A. Kasprzhitskii et al. (2022). 'Adsorption Mechanism of Aliphatic Amino Acids on Kaolinite Surfaces'. *Appl Clay Sci* 226: 106566.

large polymers. Although the linkages certainly contained more than just the canonical peptide bonds, these experiments helped galvanize efforts to better understand potentially prebiotic polymerization mechanisms.

Simply heating pure amino acid monomers in the dry state at less extreme temperatures of ~50–100 °C, however, affords little polymerization. This situation changes when certain metal salts and prebiotically plausible minerals are added to the reaction mixture. For example, when glycine is combined with **kaolinite**, a type of clay which likely was abundant in potential early Earth environments like hot springs, peptide oligomers are formed relatively readily, see Figure 7.2. In these experiments conducted at more moderate temperatures, aqueous solutions of glycine are added to the clay, and then subjected to multiple cycles of wetting and drying. Small glycine oligomers in these situations can be observed.

Another well-known protocol for amino acid polymerization driven by dry-down is known as salt-induced peptide formation. Under these conditions, large concentrations of salt are added to the initial aqueous mixture containing the amino acid monomers. Upon evaporation, the salt serves as a dehydrating agent, which promotes polymerization. Copper(II) is also added to these mixtures as a catalyst and dramatically improves the yields of oligomers. Cycles of wetting and drying are important for increasing yields, too, as redissolution likely affords opportunities for amino acids and newly formed peptides to rearrange themselves in close proximity, a circumstance which is needed for efficient amide bond formation during the dry-down stage, see Figure 7.3.

Not only do inorganic additives like clays and salts help to oligomerize amino acids by wet-dry cycles, so can other organic molecules, namely, α-hydroxyacids. These compounds can be considered as structural cousins to the amino acids, differing only by an OH group in place of the NH_2 group at the α-carbon. In fact, α-hydroxyacids and α-amino acids share common synthetic pathways (e.g. the Strecker reaction discussed in Chapter 6), and amino acids and hydroxy acids are often found together in meteorite samples.

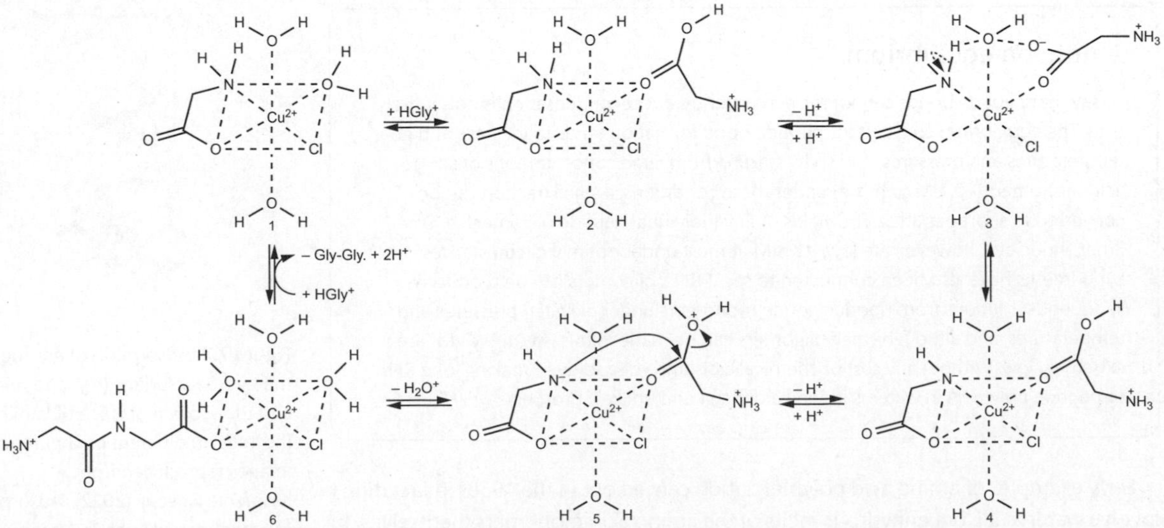

Figure 7.3 Scheme for salt-induced peptide-bond formation with copper(II) serving as a catalyst. For more details, see T. A. E. Jakschitz, B. M. Rode. (2012). 'Chemical Evolution from Simple Inorganic Compounds to Chiral Peptides'. *Chem Soc Rev* 41: 5484–5489.

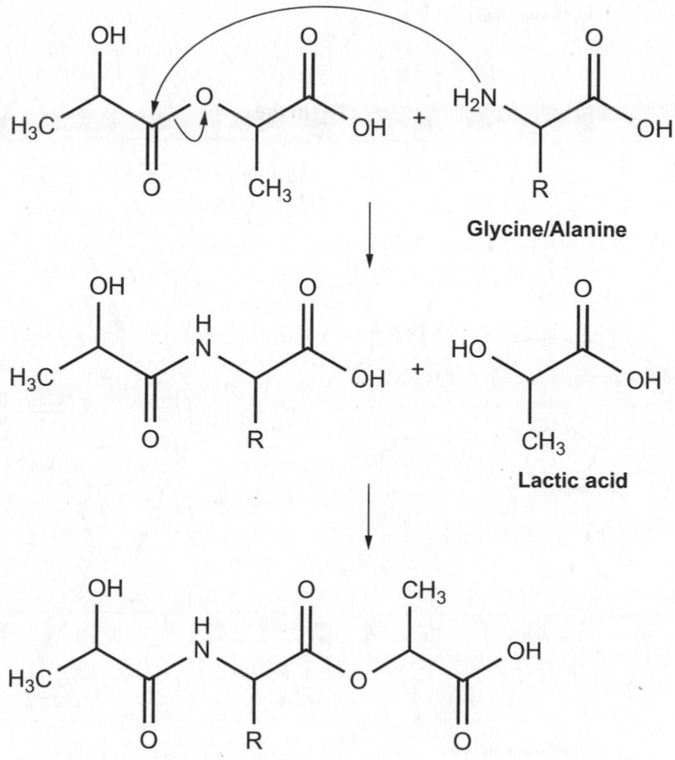

Figure 7.4 Hydroxy acid-mediated prebiotic peptide-bond formation. For more details, see J. G. Forsythe et al. (2015). 'Ester-Mediated Amide Bond Formation Driven by Wet–Dry Cycles: A Possible Path to Polypeptides on the Prebiotic Earth'. *Angew Chem Int Ed 54*: 9871–9875.

When mixtures of α-hydroxy acids and α-amino acids (for example lactic acid and glycine) are subjected to multiple cycles of dry-down, a mixture of **depsipeptides** results. A depsipeptide is a peptide in which one or more of its amide groups are replaced by esters. The mechanism for peptide bond formation actually progresses through intermediary ester bonds which form more readily during dry-down than amide bonds. The fact that esters are easier to form is likely related to their greater thermodynamic favourability characterized by a ΔG of ~0 as opposed to ~3.5 kcal mol^{-1} for an amide bond. Recall that esters themselves are still reactive and are susceptible to nucleophilic substitution at the carbonyl carbon. Intermolecular nucleophilic attack by an amino-acid amine results in peptide bond formation. Elongation of the peptide can continue to proceed in this fashion of esterification-followed-by-nucleophilic-substitution through repetitive cycles of wetting and drying, see Figure 7.4.

Abiotic polymerization of RNA is also thermodynamically unfavourable to nearly the same degree as peptides. Early attempts to make RNA from dry down of the mononucleotides revealed that making phosphoester bonds is possible using this strategy, however, achieving oligomers beyond a 2mer or 3mer has proven to be difficult. There is the added complication in such experiments that 3′-5′ phosphodiester linkages are formed alongside 2′-5′ linkages with little-to-no preference.

One strategy to overcome this limitation of dry-down that has been met with success is known as lipid-assisted synthesis. Ribonucleotides when mixed with

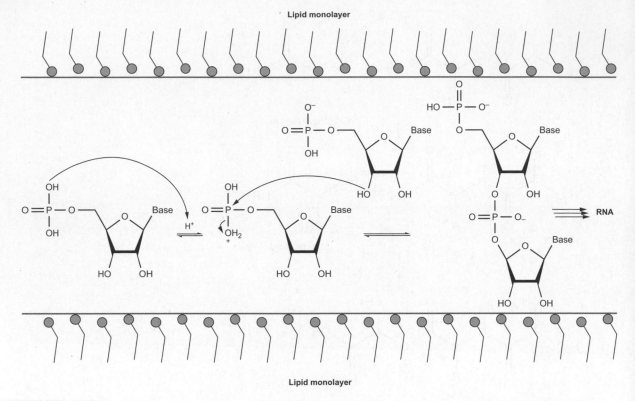

Figure 7.5 Proposed mechanism for lipid-assisted prebiotic RNA synthesis. For more details, see S. Rajamani et al. (2008). 'Lipid-Assisted Synthesis of RNA-Like Polymers from Mononucleotides'. *Orig Life Evol Biosph 38*: 57–74.

phospholipids in aqueous solution and then subjected to multiple cycles of dry-down can produce relatively long RNA-like polymers. The proposed mechanism involves the lipids in the dry state forming self-assembled lamellae (layers) that serve to organize the monomers and provide a degree of diffusional mobility that allows them to encounter each other and react. Heating provides the driving force that pushes the reaction forward by elimination of water, see Figure 7.5. Upon rewetting, the lipids spontaneously form **vesicles** with the resulting RNA-like polymers encapsulated within their interiors, hinting at a mechanism for the formation of protocells—a topic we will discuss in detail in Chapter 8.

7.2 Polymerization with Condensing Agents

For polymerization of RNA or peptides to proceed spontaneously in aqueous solutions to a significant extent under normal conditions, additional chemical energy is required. Biology is no exception to this thermodynamic constraint, whose pathways use the chemical energy stored in the phosphoanhydride bonds of nucleoside triphosphates (NTPs) to drive polymerization of RNA (not to mention a whole range of other reactions) forward. In the case of biochemical peptide synthesis, biology uses 'charged' transfer RNA (tRNA), which rely on high-energy ester bonds to render amide bond formation spontaneous. In this sense, we say these amino acids and

Table 7.1 Prebiotic condensation agents and their free energies of hydration.

Entry	Activating agent	Hydrolysis/hydration product	$\Delta G/\text{kJ mol}^{-1}$
1	NH_2CONH_2	$CO_2 + NH_3$	-16^a
2	COS (g)	$CO_2 + H_2S$	-17^a
3	Pyrophosphate	Phosphate	-19^b
4	CO (g)	HCO_2H	-16^a
5	$HNCO$	$CO_2 + NH_3$	-54^a
6	HCN	$HCO_2H + NH_3$	-75^a
7	RCN	$RCO_2H + NH_3$	-80^c
8	NH_2CN	Isourea	-83^d
9	$HNCNH$	Isourea	-97^d
10	$HCCH$ (g)	CH_3CHO	-112^a

^{a} From thermodynamic data. ^{b} Experimental determination. ^{c} Assessment from experimental and thermodynamic data. ^{d} From *ab initio* calculations. Table data reproduced from G. Danger et al. (2012). 'Pathways for the Formation and Evolution of Peptides in Prebiotic Environments'. *Chem Soc Rev 41*: 5416–5429.

ribonucleotides are *chemically activated*. Is there a way to chemically activate ribonucleotides and amino acids in a prebiotically plausible manner that makes their polymerization in water spontaneous?

The answer to this question is yes, and polymerization in aqueous solution can be made spontaneous by means of condensation agents capable of chemically activating amino acid and ribonucleotide substrates. Condensing agents provide chemical activation by installing good leaving groups on the relevant electrophilic reaction centre that requires substitution to form either the phosphoester or amide bond. The net reaction sees a new phosphoester or amide bond being formed along with concurrent hydration of the condensation agent, a stable product which ultimately provides the thermodynamic driving force for the reaction. Hence, a good condensation reagent should also be able to undergo spontaneous hydration. Table 7.1 shows a list of some prebiotically plausible condensing agents and their free energies of hydration. In the remainder of this section, we will highlight a couple of examples.

Chemical activation of amino acids

For amino acid chemical activation, the OH of the carboxylic acid must be converted to a good leaving group, which can be substituted after nucleophilic attack by the amine of another amino acid. Transforming the OH group in turn requires that the oxygen atom must first react as a nucleophile with the electrophilic centre of a condensing agent. A variety of potentially prebiotic condensing agents have been evaluated for their ability to promote peptide bond formation, all of which possess an electrophilic carbon. One notable example is carbonyl sulfide (COS), which is plausible in that it is known to be produced geochemically via volcanic outgassing. This compound is relatively efficient at facilitating peptide bond formation, especially in the presence of additional metal additives.

The mechanism is thought to proceed through a reactive intermediate known as an *N-carboxyanhydride*. The reaction mechanism begins with nucleophilic attack of the amino acid amine on the electrophilic carbon of COS. The resulting intermediate rapidly undergoes cyclization to the five-membered N-carboxyanhydride, eliminating a molecule of ^-SH. The carboxylic acid of the amino acid converted to the anhydride is now fitted with a good leaving group (albeit an intramolecular one) allowing amide bond formation to take place spontaneously. The amine of another amino acid attacks the anhydride carbon, followed by opening of the ring. At this point a dipeptide has been formed, although with a carboxyl group still on the terminal amine. The carboxyl group then eliminates as a molecule of CO_2, yielding the dipeptide with

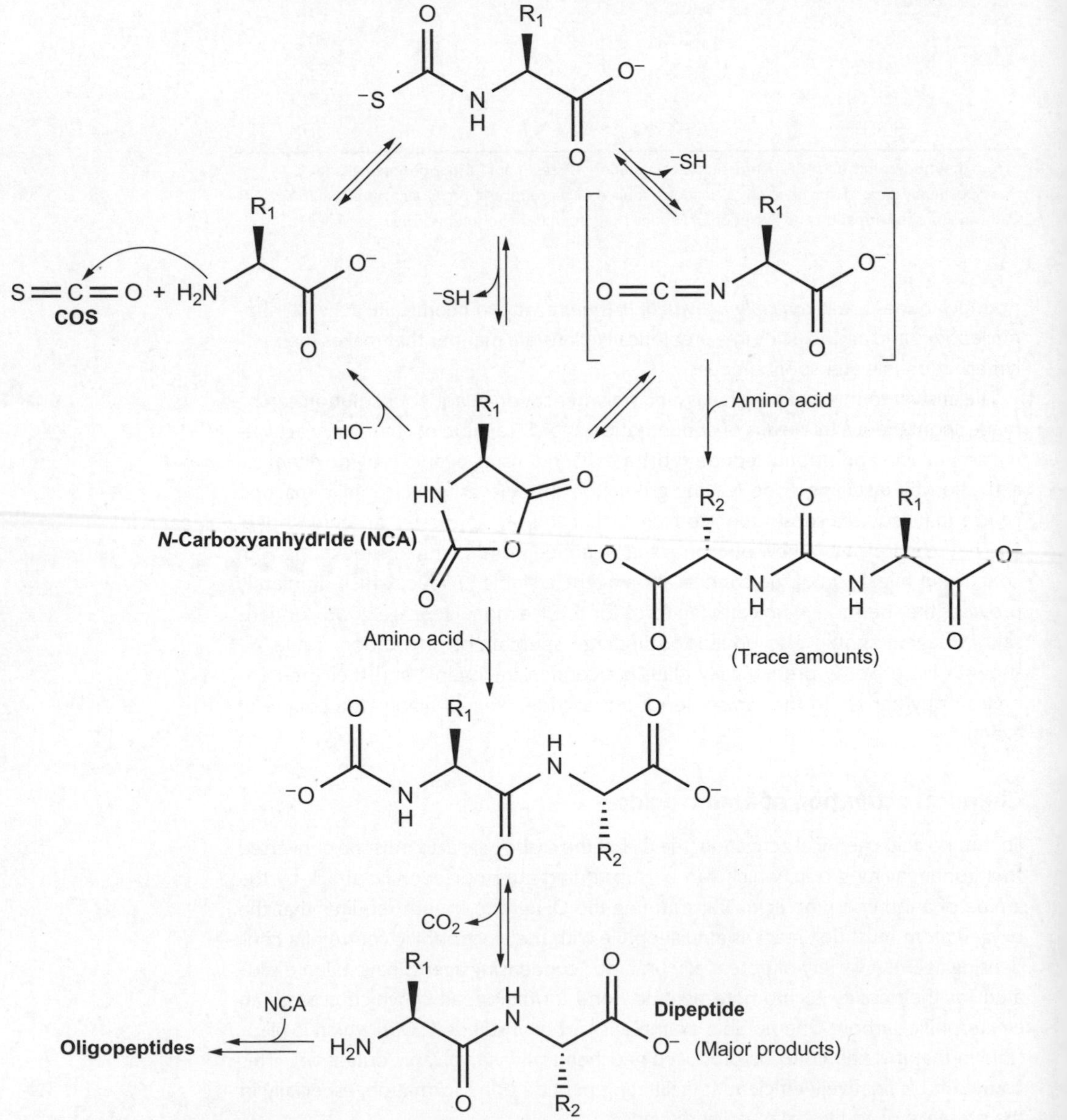

Figure 7.6 COS-mediated prebiotic peptide synthesis showing key mechanistic steps, e.g. formation of the N-carboxyanhydride intermediate. For more details, see L. Leman et al. (2004). 'Carbonyl Sulfide-Mediated Prebiotic Formation of Peptides'. *Science 306*: 283–286.

a free amine. The fact that CO_2, which is an exceptionally stable molecule as well as a gas forms in the final step helps drive this reaction forward thermodynamically. The yields of this reaction can be further increased by including metals which act as 'sponges' for HS^-, forming stable metal-sulfide bonds, or oxidizing agents which oxidize the initial N-thiocarboxy intermediate to the dimeric disulfide derivative ($-S-S-$), which is an even better leaving group than HS^- and facilitates the formation of the N-carboxyanhydride. Note, this mechanism allows for the formation of larger peptides, one amino acid at a time, see Figure 7.6.

There is another strategy for amino acid activation that takes a more bottom-up approach in the sense that the activation protocol begins from α-aminonitrile precursors of amino acids rather than the carboxylic acids themselves. This protocol requires that the N-terminus of the growing peptide chain be protected by an acetyl group, which is achieved by acylation of the initial α-aminonitrile. Next, the nitrile of this protected compound is converted to a thioacid by reaction with ^-SH followed by hydrolysis of the thioamide intermediate. The SH functionality can then be transformed into a good leaving group, like we saw in the previous example, by oxidation to the dimeric disulfide derivative, or by using a number of other prebiotically plausible condensing agents with electrophilic carbons that further activate the SH group. Substitution at the carbonyl carbon with another α-aminonitrile allows for the cycle of elongation to be repeated, see Figure 7.7.

Chemical activation of ribonucleotides

Because of the difficulties of polymerizing RNA from mononucleotides in the dry state, not to mention in aqueous solutions, the majority of prebiotic RNA polymerization research has focused on the use of prebiotically plausible chemical activation.

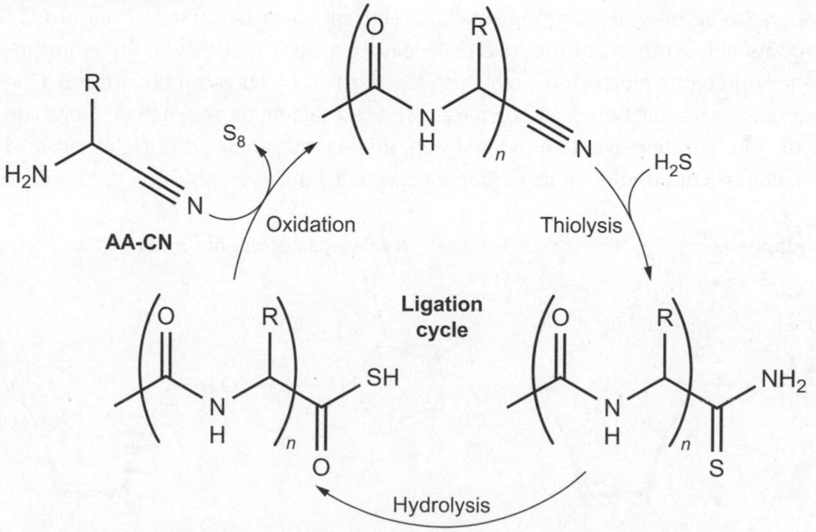

Figure 7.7 Aminonitrile thiolysis-mediated prebiotic peptide synthesis showing key mechanistic steps. For more details, see P. Canavelli et al. (2019). 'Peptide Ligation by Chemoselective Aminonitrile Coupling in Water'. *Nature* 571: 546–549.

In the case of RNA, the electrophilic reaction centre is the phosphorus atom of the organic phosphate group. The most obvious choice for prebiotic RNA polymerization is to use NTPs, from the point of view that they are what extant biology already uses (Figure 7.8). The NTPs can be thought of as nucleoside monophosphates installed with a good leaving group in the form of pyrophosphate, and there are potentially prebiotic mechanisms by which the NTPs can be synthesized. In the process of forming a new phosphoester bond, pyrophosphate is eliminated, a by-product which can be further hydrolysed to two molecules of phosphate providing a further thermodynamic driving force for polymerization. Although the formation of phosphodiester bonds starting from NTPs is thermodynamically favourable, the kinetics of these reactions without enzymes are so slow that hydrolysis of the phosphoester bonds competes with their formation when relying on these substrates. Nevertheless, it is possible to make RNA prebiotically using the NTPs when dry-down steps are included which speed up the kinetics. Recently, it has been shown that some types of basaltic glasses, which are a type of abundant volcanic rock composed mostly of silica, are able to accelerate dramatically the polymerization of the NTPs when they are adsorbed to their surfaces.

One of the most well-studied potentially prebiotic leaving groups for facile nonenzymatic RNA polymerization are the imidazoles. Ribonucleotides chemically activated with imidazole derivatives are known as **phosphorimidazolides**, chemically activated compounds in which the imidazole is bonded to the phosphorus atom through a P–N linkage, replacing one of the phosphate OH groups (Figure 7.8). Imidazoles are excellent leaving groups that are kinetically labile enough to allow phosphodiester bond formation to occur on timescales of minutes to hours. These fast kinetics allow nonenzymatic RNA synthesis to be studied on convenient laboratory timescales, thus serving as a useful model for prebiotic chemistry, but whether phosphorimidazolides would have been present on early Earth is still debated. Nevertheless, potentially prebiotic syntheses of some imidazole derivatives have been reported, in particular, 2-aminoimidazole which is among the most efficient imidazole-based leaving groups.

Along with synthesis of the imidazole leaving groups themselves, there are potentially prebiotic mechanisms by which these leaving groups can be installed. One such chemical activation protocol makes use of acetaldehyde and methyl isocyanide as the condensation agent. In the first step, methyl isocyanide adds to the carbonyl of acetaldehyde, rendering its carbon electrophilic and susceptible to nucleophilic

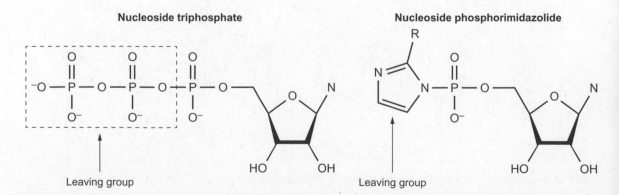

Figure 7.8 Chemical structures of a generic nucleoside trisphosphate and nucleoside phosphorimidazolide.

attack by the phosphate of a nucleotide. The resulting imidoyl phosphate-bearing nucleotide is now installed with a good leaving group which can be substituted by imidazole derivatives, eliminating a hydrated amide by-product in the process. Phosphorimidazolides are known to oligomerize in aqueous solutions given the presence of an appropriate divalent metal catalyst (e.g. Mg^{2+}), but when they are adsorbed to the surface of **montmorillonite**, a type of clay mineral, even longer RNA polymers can be produced, see Figure 7.9.

A)

B)

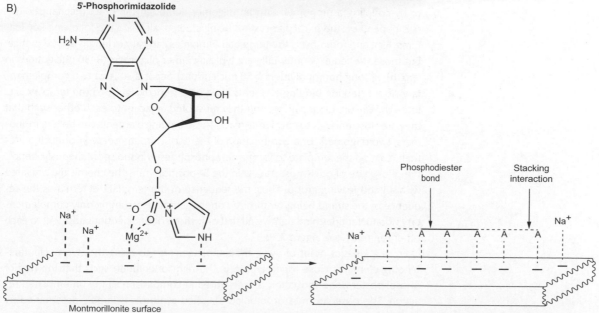

Figure 7.9 A) Image of montmorillonite, a type of clay mineral. **B**) Imidazole-activated RNA polymerization mediated by adsorption onto montmorillonite. Scheme reproduced from K. Kawamura, J. P. Ferris. (1994). 'Kinetic and Mechanistic Analysis of Dinucleotide and Oligonucleotide Formation from the 5'-Phosphorimidazolide of Adenosine on Na^+-Montmorillonite'. *J Am Chem Soc 116*: 7564–7572.

7.3 Nonenzymatic Template-Directed Synthesis

In 1953, Watson and Crick published their ground-breaking structure of the DNA double helix based on the X-ray data gathered by Rosiland Franklin and Maurice Wilkins, and in the final paragraph of their article, they famously wrote, 'It has not escaped our notice that the specific [Watson–Crick] pairing we have postulated immediately suggests a possible copying mechanism for the genetic material'. Indeed, the ability of the nucleobases to recognize specific and complementary base-pairing partners is the fundamental basis through which **template-directed synthesis** of nucleic acids can occur, a mechanistic feature that maintains a central role in both extant biology and the hypothetical RNA world.

Nonenzymatic template-directed synthesis is the postulated means by which RNA, or perhaps another structurally similar polymer that arose geochemically, originally copied itself chemically before the aid of enzymes, or even the advent of any early ribozymes. In essence, nonenzymatic template-directed synthesis is thought to be a possible mechanism that 'jump-started' Darwinian evolution. Although much less researched in comparison, nonenzymatic template-directed synthesis of peptides is also necessary if peptides are to have played an inheritable role in prebiotic metabolism prior to the emergence of the ribosome predecessor capable of producing peptides of arbitrary sequence. In this last section of the chapter, we'll discuss mechanisms for both nonenzymatic template-directed RNA and peptide synthesis.

General mechanistic features for RNA template-directed synthesis

Nonenzymatic template-directed RNA synthesis can take place through on-template polymerization of monomeric ribonucleotides, ligation of oligomer ribonucleotides, or by coupling a monomer with an oligomer. Although the exact mechanistic details depend on the particular system being studied, all share the following key features. First and foremost is the template. Templates can vary in length and sequence, but those studied experimentally are typically either oligomers (8~30 nucleotides in length), or long homopolymers (>50 nucleotides). Second, at least two complementary ribonucleotide binding elements must bind noncovalently to the template, i.e. through Watson–Crick pairing, and they must bind adjacent to each other such that they are close enough to covalently react. These binding elements can be two monomers, two oligomers, or a combination of the two. Third, these two ribonucleotides then react on the template to form a new phosphoester bond spontaneously; hence, the phosphate of one (most usually in the 5'-position) must be chemically activated with a good leaving group. Thus, the sequence of the template determines the sequence of the strand being synthesized on the template assuming only complementary ribonucleotides bind tightly and that only those ribonucleotides adjacent to each other can react, see Figure 7.10.

From a kinetics point of view, the template serves to bring the two nucleotide-binding elements close together and in the right orientation such that the rate of phosphoester bond formation is accelerated in comparison to the untemplated reaction. The observed rate of template-directed synthesis is largely determined by the binding constants of the nucleotides to the template and the rate constant for on-template phosphodiester bond formation.

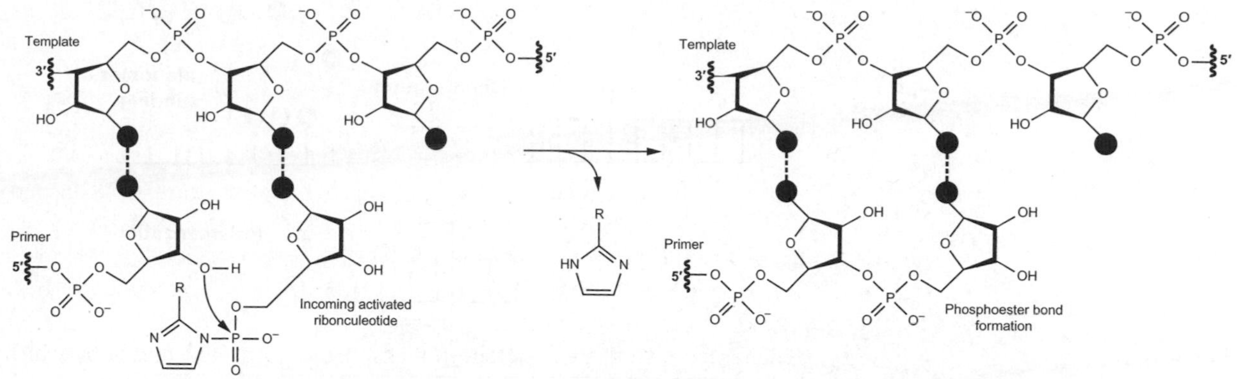

Figure 7.10 General mechanism for nonenzymatic template-directed RNA synthesis relying on activated 5′-phorphorimidazolide ribonucleotides.

Like a catalyst, the template serves to increase the rate of reaction between the (activated) ribonucleotides, which are complementary to its sequence. But a template strictly speaking is not a true catalyst, because the template will be consumed by the end of the reaction, usually in the form of a stable double helix. To be a true catalyst, the template must be liberated from the newly synthesized complementary strand at the end of the synthesis. For long duplexes, this turns out to be a nontrivial issue as a consequence of the fact that duplex stability generally scales with length. Nevertheless, assuming the two strands can be separated, then the door is also open for autocatalysis, that is, if the newly synthesized complement strand can itself be used as a template for the synthesis of the original template strand, see Figure 7.11.

Another important mechanistic concept in template-directed synthesis is **fidelity of copying**. In an ideal world, only ribonucleotides complementary to the template will bind and react with each other, while reactions involving noncomplementary nucleotides never occur. Of course, reality is more complicated than this ideal situation, and noncomplementary ribonucleotides from time-to-time also incorporate themselves into the sequence. Nonstandard base-pairing, although typically weaker than Watson–Crick pairing, can still occur and lead to 'incorrect' sequences. Fidelity is defined as the fraction of correct nucleotide incorporation with respect to incorrect nucleotide incorporation. For example, if G is the next nucleobase in the template sequence to be copied, then the fidelity of this reaction is the amount of complementary C incorporated divided by the total amount of A, U, G, and C incorporated. Hence, the fidelity can also be defined in terms of the observed rates of each of these reactions, i.e. the rate of C incorporation divided by the total rate for all.

Strategies for on-template RNA polymerization and ligation

Early demonstrations of template-directed RNA polymerization used homopolymeric templates and activated nucleotide monomers. In particular, it was discovered that the identity of the leaving group has a significant influence on the reaction outcome. Ribonucleotides activated with 2-methylimidazole (as opposed to just imidazole) proved to be especially good for template-directed synthesis. In particular, with 2-methylimidazole as the leaving group, heteropolymers with all four nucleobases

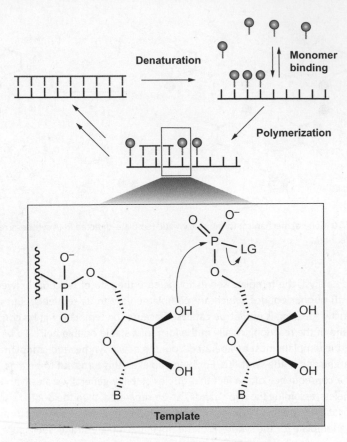

Figure 7.11 General scheme for achieving multiple rounds of nonenzymatic template-directed RNA synthesis via duplex strand separation (denaturation).

could be copied, so long as these templates contained at least 60% of C nucleobases. In the proposed mechanism, the imidazole of the phosphorimidazolide first becomes protonated giving it a positive charge before substitution can occur. This mechanism suggests that *N*-alkylated phosphorimidazolides, which bear a 'permanent' positive charge should be excellent leaving groups, and indeed they are despite whether they are prebiotically plausible or not.

Although a success, one of the issues discovered early on is that some nucleobases are more easily incorporated than others. Incorporation of G opposite to C in the template strand is the most efficient, while incorporation of U opposite to A is the least. At least part of the reason is the fact that uridine monophosphate has the weakest binding strength. In fact, having two A nucleobases in a row in the template strand almost completely inhibits further template-directed synthesis.

One strategy to help increase the binding constants of the mononucleotides, especially U, is to use 'helper' oligonucleotides. These helpers are usually a few nucleotides in length and bind just downstream of the mononucleotide binding site. The extra base-stacking surface these helper oligonucleotides provide can increase the binding constants of the mononucleotides by approximately an order of magnitude,

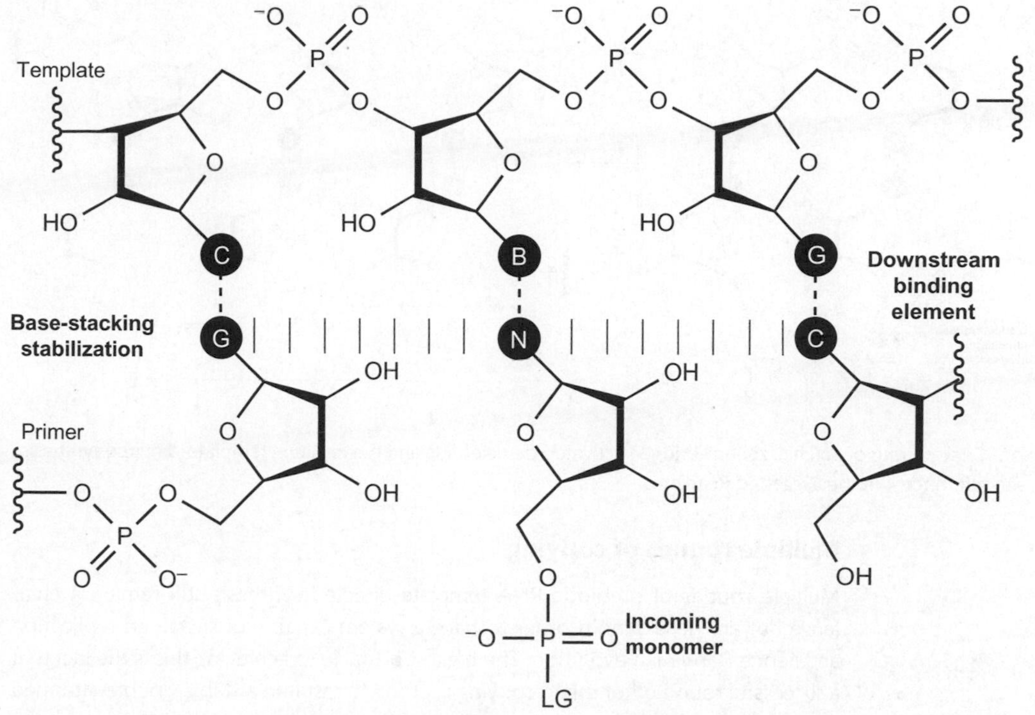

Figure 7.12 Downstream helper oligonucleotides increase the binding affinity of activated ribonucleotides through additional base-stacking stabilization. For more details, see reference 4 in the Suggested Reading.

see Figure 7.12. In fact, it is base-stacking rather than hydrogen bonding that provides the majority of duplex stability.

Another strategy to help mitigate this inefficiency of U incorporation is to use activated oligonucleotides instead of mononucleotides and to carry out template-directed ligation reactions. Oligonucleotides even only as long as di- or trinucleotides can bind with significantly more affinity and have the added benefit that they can also potentially lead to higher-fidelity copying. They can also potentially copy over RNA sequences that have secondary structure, e.g. hairpin stem loops (see Chapter 5).

Another important mechanism for efficient nucleotide incorporation was discovered somewhat serendipitously and involves imidazolium-bridged dinucleotides. Two phosphorimidazolides can react intermolecularly, with the imidazole N atom of one, nucleophillically substituting the imidazole of the other. The resulting imidazolium-bridged structure bearing a positive charge possesses an excellent leaving group in the form of an intact phosphorimidazolide, while the dinucleotide structure affords increased affinity to the template through two Watson–Crick base pairs. These bridged-imidazolium dinucleotides react an order of magnitude faster in comparison to the monomer phosphorimidazolides from which they are derived, see Figure 7.13.

In terms of increasing the rate of the on-template covalent reaction, one strategy is to enhance the nucleophilicity of the elongating strand by swapping out the 3'-OH for an NH_2 on the ribose component. Although not a canonical RNA structure, the rates of on-template copying are dramatically increased.

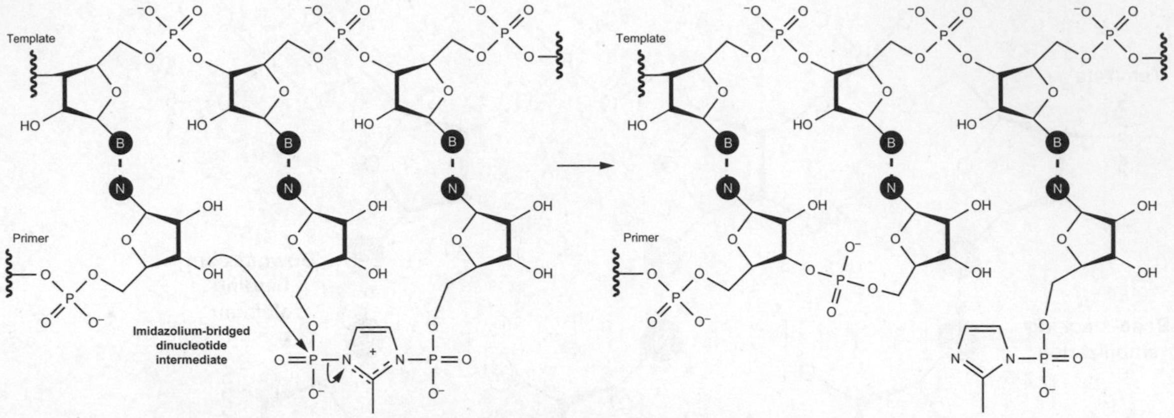

Figure 7.13 Watson–Crick pairing of an imidazolium-bridged activated dinucleotide and the resulting template-directed synthesis. For more details, see reference 4 in the Suggested Reading.

Multiple rounds of copying

Multiple rounds of prebiotic RNA template-directed synthesis still remain a challenge but are necessary in order to have a system capable of sustained replication and hence Darwinian evolution. The biggest obstacle to achieving this is the fact that a successful round of template copying is prone to result in a stable double-stranded structure. In order for another round of template-directed synthesis to begin, the two strands must be separated and remain separated long enough for on-template polymerization/ligation to take place. The fact that the resulting duplex impedes further rounds of copying is known as the **strand inhibition problem**. Two hybridized strands can potentially be separated simply by raising the temperature past their melting points; however, high temperatures also prevent Watson–Crick pairing of the activated mono- and oligonucleotides.

One strategy is to carry out the synthesis in a viscous gel-like fluid. Upon heating, the two strands can separate, but upon cooling back down, the viscous medium prevents the relatively large strands from quickly re-annealing to such a degree that template-directed copying can take place, see Figure 7.14.

Another rather ingenious strategy for achieving multiple rounds of RNA synthesis takes advantage of **chimeric templates**, which are sequences composed of two different types of nucleotides, for example, both DNA and RNA. These chimeric sequences are still capable of binding pure RNA oligonucleotides affording their template-directed ligation; however, the resulting RNA-chimeric duplex is not as stable as the corresponding pure RNA duplex. This weaker duplex means the chimeric strand can be readily displaced by two, smaller complementary RNA oligonucleotides, which then undergo template-directed ligation on the newly formed RNA strand. Thus, the chimeric template is released at the end of the reaction acting as a true catalyst that can afford multiple rounds of template-directed RNA synthesis, see Figure 7.15.

Achieving multiple rounds of RNA copying still has challenges to be overcome but is currently an active area of research. Success would open the door to producing chemical systems which can begin to undergo Darwinian evolution in the laboratory in a potentially prebiotic manner.

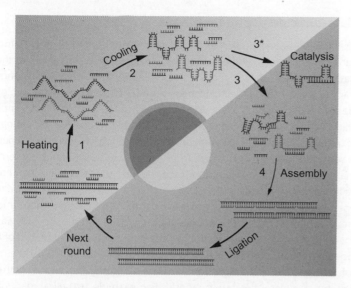

Figure 7.14 Scheme showing strand re-annealing inhibited by viscous gels affording multiple rounds of nonenzymatic template-directed synthesis. Scheme reproduced from C. He et al. (2019). 'Solvent Viscosity Facilitates Replication and Ribozyme Catalysis from an RNA Duplex in a Model Prebiotic Process'. *Nucleic Acid Res* 47: 6569–6577.

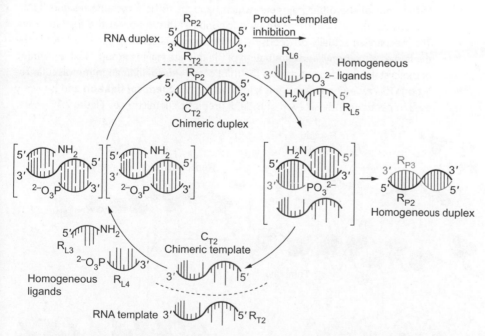

Figure 7.15 Scheme showing multiple rounds of nonenzymatic template-directed synthesis using a chimeric template composed of both DNA and RNA nucleotides acting as a catalyst. Scheme reproduced from S. Bhowmik, R. Krishnamurthy. (2019). 'The Role of Sugar-Backbone Heterogeneity and Chimeras in the Simultaneous Emergence of RNA and DNA'. *Nature Chem* 11: 1009–1018.

Template-directed peptide synthesis

Peptides have long been studied in the context of prebiotic chemistry as there are facile pathways for abiotic synthesis of amino acids (see Chapter 6) as well as their polymerization. The rich catalytic and structural roles of peptides are limited, however, by the lack of a generalized mechanism for peptide self-replication and evolution.

Unlike nucleotides, amino acids do not have specific, complementary molecular recognition partners between themselves, like Watson–Crick pairing, rendering a generalized template-directed mechanism difficult to envision. Hence, even if a particular peptide sequence happened to confer an advantage to an early life-like system, there is no analogous universal mechanism for that peptide to replicate itself and thereby become heritable. The RNA world hypothesis offers a solution in that peptide replication depended on and occurred after the evolution of RNA catalysts (ribozymes), but the general consensus that peptides were almost certainly present on early Earth has led to investigations of the potential for peptide replication without the aid of RNA.

Peptides have the potential to display secondary/tertiary structures that are capable of well-understood recognition and **self-assembly** behaviours that can be exploited for template-directed synthesis. For example, α-helical peptides can serve as a template for their own ligation products from shorter peptide oligomers. The template α-helical peptide in this case first binds with the shorter oligomers that together make up the same sequence as the template strand. One side of the α-helical template is decorated with hydrophobic residues, which provide the basis for recognition of two of the smaller fragments. The C-terminus of one of the fragments is chemically activated as a thiolate ester. Once both are bound to the template, the N-terminus of the other fragment, which happens to be a cysteine residue, spontaneously forms a peptide bond by nucleophilic substitution at the thiolate ester. The mechanism actually proceeds by nucleophilic attack of the cysteine thiol at the thiolate ester carbonyl and elimination of the thiolate leaving group. Next, the amine of the cysteine residue substitutes the thiol of the sidechain in an intramolecular reaction. This type of mechanism is known as **native chemical ligation** and is widely used in peptide synthesis even outside of prebiotic chemistry, see Figure 7.16.

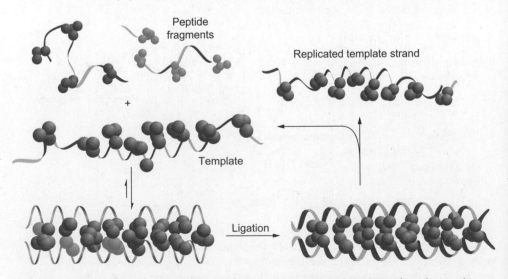

Figure 7.16 Nonenzymatic template-directed peptide ligation using an α-helical peptide as the template. For more details, see D. H. Lee et al. (1996). 'A Self-Replicating Peptide'. *Nature 382*: 525–528.

There are many other examples of template-directed peptide synthesis that follow this type of mechanism, but more research is needed to understand if these types of protocols could have provided a means for heritable peptides during the early stages of prebiotic chemistry.

7.4 Summary

- The two main strategies to achieve amino acid and ribonucleotide oligomers/polymers are (*i*) cycles of wetting and drying and (*ii*) chemical activation with condensation agents.

- RNA may have originally been copied via nonenzymatic template-directed synthesis before the advent of enzymes.

- Multiple rounds of prebiotic RNA template-directed synthesis are necessary to have a system capable of sustained replication. One obstacle to achieving this is the stability of the double helix, known as the *strand inhibition problem*.

- Peptides can display secondary/tertiary structures that are capable of recognition and self-assembly and can undergo template-directed synthesis.

7.5 Exercises

1. Is the imidazole in the phosphorimidazolide shown in Figure 7.9 in its current form a good leaving group? Why or why not?

2. Imagine you were to carry out a nonenzymatic copying reaction on an RNA that was also a ribozyme, which takes on a stable, folded three-dimensional (3D) structure. What strategies might you use to help ensure the copying reaction takes place smoothly?

3. A catalyst accelerates the rate of a reaction by lowering the energy of the transition state. Does a template also accelerate the rate of a reaction in the same way?

4. Describe a protocol which you might use to achieve multiple rounds of nonenzymatic template-directed RNA synthesis.

5. Describe a way in which you can use peptide secondary structures as recognition motifs to enable nonenzymatic template-directed peptide synthesis.

7.6 Suggested Reading

1. Jack W. Szostak. (2012). 'The Eightfold Path to Non-Enzymatic RNA Replication'. *J Sys Chem 3*: 2.

2. Bruce Damer and David Deamer. (2020). 'The Hot Spring Hypothesis for an Origin of Life'. *Astrobiol 20* (4): 429–452.

3. Orgel Leslie E. (2004). 'Prebiotic Chemistry and the Origin of the RNA World'. *Crit Rev Biochem Mol Biol 39*(2): 99–123.

4. Marilyne Sosson and Clemens Richert. (2018). 'Enzyme-Free Genetic Copying of DNA and RNA Sequences'. *Beilstein J Org Chem 14*: 603–617.

5. James P. Ferris. (2006). 'Montmorillonite-Catalysed Formation of RNA Oligomers: The Possible Role of Catalysis in the Origins of Life'. *Phil Trans R Soc B 361*: 1777–1786.

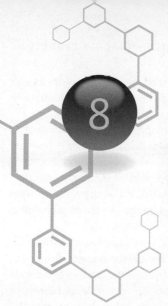

8 Protocells: Compartmentalization, Replication, and Integrated Molecular Function

While all contemporary life is cellular, it is not clear how life became this way. The concept of the protocell has been introduced to scaffold thinking on what intermediary types of material organization may have existed during the evolution of life. Protocells in this context are thus conceptual molecular assemblages which provide a defining boundary for chemistry which occurs during their collective reproduction process, with varying capacities for fulfilling the other signature functions of living systems such as genetic information transfer or the ability to undergo Darwinian evolution. At some point protocells are hypothesized to have been able to replicate all of their internal informational content, while making use of chemical 'nutrients' provided by the environment. The importance of compartments and different types of compartment materials including lipid membranes is discussed here. Current models for protocells based on ribonucleic acid (RNA), fatty acid vesicles, and small peptides are presented. Possible mechanisms of protocell reproduction which entail RNA replication (i.e. nonenzymatic template-directed synthesis) and membrane division are highlighted. The concepts of a minimum cell and synthetic biology are also introduced.

8.1 Protocells and the Importance of Compartments

A protocell is any theoretical or experimental model involving a cell-like compartment, e.g. a bounded structure which impedes the free flow of material between the interior and exterior of the structure, in which chemical processes take place inside of it and/or in the surrounding environment and that help mimic or explain the origin of more complex biological cellular functions. Protocells generally serve as models for understanding how the phenomenon of encapsulation or cellularization affected the development of interior cell chemistry, and integrated cellular functions. Protocell compartments made in the lab are often constructed from phospholipids or fatty acids, which mimic the structure and function of extant biological membranes, and whose self-assembly properties are predictable, but other types of compartments have also been investigated. The term 'protocell' can also refer to the hypothetical primitive cells that existed on early Earth during the era of prebiotic chemistry, as well as to laboratory models of contemporary cell functions, which do not intentionally consider the origins or evolution of life.

Compartmentalization is evidently essential to biology, and a crucial concept in the story of life's origins. All contemporary life is cellular, and even multi-cellular organisms are compartmentalized at other hierarchical levels. Compartmentalization is essential for giving individual organisms their identity. Could life exist without a compartment of some sort? What is it about compartments that makes them so necessary for life? Do compartments need to be based on amphiphiles like the phospholipids that extant biology employs?

Compartments are necessary for life as we know it, partly because they allow for the Darwinian evolution of complex adaptive traits. Consider for example, that prebiotic chemistry at some point produced a catalytic RNA oligomer capable of replicating other RNA molecules (including itself) making use of nucleoside triphosphates (NTPs) as substrates. Imagine this RNA replicase, free in solution, encounters by chance another catalytic RNA, i.e. an RNA NTP-synthase, capable of producing NTPs from the available abiotic feedstocks. The RNA replicase then copies this RNA NTP-synthase, but without any compartmentalization, the NTP-synthase diffuses away, generating NTPs elsewhere at a distant location too far to assist the RNA replicase in replicating any more RNAs. A similar situation is true for the RNA NTP-synthase—it and its NTPs are too far away from the replicase such that it can't be copied again and is eventually hydrolysed to the point of disfunction. While it may be possible for some relatively simple functional traits to evolve, compartmentalization is needed to gather functional reactions into a single small volume, such that they can behave as a single system that shares the benefits of their collective function as one evolutionary unit. Only then can more sophisticated, collective, and evolvable behaviours start to emerge. (For further discussion, see G. F. Joyce, J. W. Szostak. (2018). 'Protocells and RNA Self-Replication'. *Cold Spring Harb Perspect Biol 10*: a034801.)

At a more abstract level, compartmentalization allows cells as a population to tinker with developing internal rules which do not apply in general in the surrounding environment. This means each internal compartment becomes a finite playground for the almost infinite combinatorial possibilities of organic chemistry. Defined and reinforceable feedbacks among small molecules become possible. Among combinatorial networks of chemical reactions, some common processes may be common only within compartments, for example, interconnected autocatalytic systems amplified by weakly associated feedback processes. (Recall the chemoton model discussed in Chapter 4.) Hence, it may be only *within* compartments that true life-like chemistry can start to be observed.

8.2 Different Types of Protocell Compartments

It is not yet known whether life and its origins require lipid-based membrane compartments, or if other types of materials/interface boundaries are also sufficient. The origin of life may have initially depended on alternative compartment types that later evolved in to phospholipid compartments. It is possible the first cell-like containers could have used other types of amphiphiles besides phospholipids or have been based on peptides or some other organic structure entirely. The first compartments may even have been based on inorganic materials, like porous rocks. Whether the first biocompartments were lipid- or protein-based, or based on inorganic confinement or some combination of these remains an open question.

Compartments often selectively limit diffusion across their boundaries, and in principle, diffusion can be restricted even without clearly defined physical boundaries. For example, molecules adsorbed on a mineral surface have restricted diffusion, and so this confinement in a sense is a type of compartmentalization albeit two-dimensional (2D). In this section, we discuss different types of compartments that have been proposed as important in various origins-of-life scenarios. We restrict our discussion, however, to those compartments clearly delineated by some sort of three-dimensional (3D) physical boundary into well-defined individual units, even if these compartments may be porous and interconnected.

Inorganic compartments

The prebiotic chemistry leading to the first protocells may have been significantly influenced by inorganic compartments, in particular, the microenvironments provided by porous rocks and minerals, the eutectic phases of ice, and atmospheric aerosol particles.

Hydrothermal vent chimneys are often composed of rocks with interconnected porous structures. It is possible these pores could have functioned as protocellular compartments, each one housing unique sets of molecules and reaction networks driven by the natural chemistry of the vents themselves, see Figure 8.1.

Eutectic phases of ice form when aqueous solutions start to freeze. The growing ice crystals exclude salts and other solutes into a concentrated fluid known as a **eutectic phase**, which towards the end of the freezing process becomes separated into disconnected microscopic fluid inclusions in the ice that could serve as compartments. Eutectic ice phases have been demonstrated to facilitate nonenzymatic

500mm

Figure 8.1 Photographic image showing the porous nature of hydrothermal vent chimneys. Image reproduced from the National Oceanic and Atmospheric Administration.

Aqueous phase   Eutectic phase

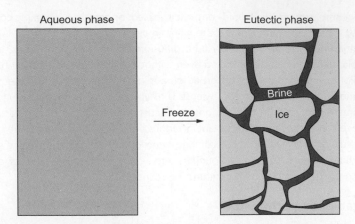

Freeze

Figure 8.2 Scheme showing the formation and general structure of a eutectic ice phase.

polymerization of initially dilute activated nucleotides due to the extreme increases in concentration they facilitate, see Figure 8.2.

Aerosols are microdroplets of water (or solid particles covered by water) suspended in the atmosphere, that can form from wind action on bodies of water. For example, think of the sea spray created by crashing waves. These droplets can remain suspended in the air for significant amounts of time during which they often lose most of their initial water content. Such aerosols could have served as compartments that afforded unique chemical reactions not possible in bulk aqueous solutions and contributed significantly to the prebiotic organic inventory, see Figure 8.3.

Figure 8.3 Aerosol formation in the form of sea spray generated by wave action.

Coacervates

Coacervates are condensed droplets formed by a process known as liquid-liquid phase separation. Coacervates typically form when two oppositely charged polyvalent ions, e.g. ones with multiple positively charged sidechains, such as RNA and peptides, are mixed together in aqueous solution. Their mutual Coulombic attraction leads to aggregation into viscous, liquid microdroplets that are separate from the rest of the aqueous solution, although they have no membrane envelope. Coacervates are generally dynamic structures, which can exchange material with the surrounding environment, sequester other highly charged or polar species, and facilitate organic reactions within their interiors, including enzyme and ribozyme catalysis. Similar types of phase-separated structures involving protein aggregates are observed in cells today, and for this reason, they are sometimes referred to as 'membraneless organelles', see Figure 8.4.

Peptide-based compartments

While all biological membranes have some peptidic content, and many bacterial cells have cell walls which are entirely composed of proteinaceous material, some peptides by themselves are capable of self-assembling to form compartments, by either self-assembly into membranes or liquid-liquid phase separation into coacervates. Amphiphilic peptides composed of a hydrophilic region (e.g. a glutamate-rich sequence) and a hydrophobic region (e.g. a phenylalanine-rich sequence), can self-assemble into spherical vesicles delineated by a peptide-membrane bilayer. Coacervation can also form via spontaneous liquid-liquid phase separation from relatively simple, neutral dipeptide-derived compounds that possess appropriately hydrophobic sidechains. Relatively small peptides can even self-assemble into hollow nanocapsules which are capable of encapsulating nucleic acids, mimicking viral capsids.

Lipid-based compartments

Compartments built from lipid membrane vesicles are by far the most studied in the context of model protocell systems, since these types of bilayers most closely resemble modern cell membranes, i.e. phospholipid membranes (although it is worth mentioning that most prokaryotes have some sort of outer cell wall, which early protocells

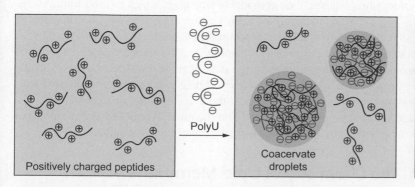

Figure 8.4 Scheme showing coacervate formation from RNA/peptide mixtures. On the right is a microscopic image of coacervate droplets. Scheme and image reproduced from W. M. Aumiller, C. D. Keating. (2016). 'Phosphorylation-Mediated RNA/Peptide Complex Coacervation as a Model for Intracellular Liquid Organelles.' *Nature Chem 8*: 129–137.

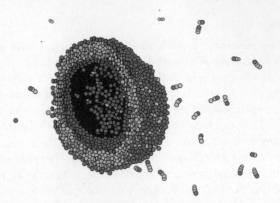

Figure 8.5 Intermediate stage in the self-assembly of lipids into a bilayer vesicle.

are typically assumed to have lacked, though it may be the case that membranes have always been reinforced by other compounds). Modern phospholipids, when isolated in the lab, can spontaneously self-assemble into vesicles under a wide range of conditions, see Figure 8.5. It is worth noting that life frequently uses molecules with a propensity to self-assemble, likely because this offers an opportunity to maximize functionality with minimal encoding.

The majority of prebiotic protocell research has focused on compartments constructed from bacterial-type as opposed to archaeal-type phospholipids and components thereof, and there could be significant nuance to how these different bilayers affect systemic function. Modern bacterial (and eukaryotic) cell membranes are largely composed of glycerophospholipids with two fatty acids esterified to glycerol with an additional ester-linked phosphate (see Chapter 3 for more details). The phosphate head group itself can be linked to other hydrophilic moieties, for example serine or choline. Archaebacteria also use glycerol as a lipid scaffold, but for the organization of ether linked isoprenoid lipids, which are similarly hydrophobic as fatty acids, but provide different integration environments for transmembrane proteins. Isoprenoid lipids have been much less investigated in the context of protocells, and it is thus possible still other types of lipids which might have scaffolded cellular evolution in important ways remain to be investigated.

Besides phospholipids, their component fatty acids, which can also form vesicles under appropriate conditions, have been a major target of investigation in protocell research. It is thought that the fatty acid components of phospholipids would have been initially produced by prebiotic chemistry, and the more complex phospholipids then emerged at a later, more advanced stage. Fatty acid vesicles also have a couple of advantages over phospholipids in terms of their physical properties that are considered necessary for early protocells, without any enzymes or sophisticated informational polymers, to have possessed. The chemistry of fatty acid and phospholipid vesicles and their physical properties are the topic of the next section.

8.3 The Chemistry of Lipid Membrane Vesicles

Vesicles are supramolecular assemblies containing an aqueous interior separated from the bulk solution by one or more bilayers composed of some type of amphiphile or mixtures of amphiphiles. Under the right conditions, vesicles form spontaneously as a result of the hydrophobic effect, which in the case of phospholipids and fatty acids, arises from the differential interactions of the hydrocarbon tails and hydrophilic head groups with the bulk solvent, i.e. water. The hydrophobic portions of amphiphilic molecules tend to aggregate in order to minimize interaction with water, while the hydrophilic portions of the molecules preferentially minimize their interactions with the surrounding polar solvent.

Vesicle self-assembly is a highly cooperative process, meaning that above a certain threshold concentration of amphiphiles in solution, known as the critical aggregation concentration (CAC), vesicle structures begin to form, while below the CAC, few-to-no vesicles exist. The ability to form vesicles also is dependent on the identity of the head group and structure and length of the hydrocarbon tail(s), in which a minimum tail length is necessary. In the case of fatty acid vesicles, for example, the minimum tail length is around eight carbons. The concept of the critical packing parameter, which describes the ratio of the volumes of the hydrophobic portion of amphiphiles to the cross-sectional area of the hydrophilic head group has been developed to explain the propensity of amphiphiles to form vesicles in solution, but it is clear this model is not sufficient to describe all possible interactional phenomenon which result in compartment formation, because such self-assembly depends on the delicate balance of many nanometer-scale physical forces.

Just like micelles, the hydrocarbon chains render the inside of the bilayer a hydrophobic environment. This nonpolar, hydrophobic region is what creates a barrier to diffusion of charged molecules across the membrane. On the other hand, the inside of the membrane can also sequester hydrophobic, nonpolar molecules, which can potentially stabilize the membrane, exhibit catalytic functions or possess photochemical activity, to name some examples.

Comparisons between fatty acid and phospholipid vesicles

In general, biologically derived double-chain phospholipids form more stable vesicles than their single-chain fatty acid counterparts. Although beneficial for extant biology, this greater stability could pose problems for hypothetical early protocells. Phospholipid membranes are practically impermeable to charged compounds. Biology has solved this problem by employing a number of active and passive ion transport proteins integrated into cell membranes, which early protocells would have lacked. This lack of permeability is especially problematic if passive diffusion were the primary mechanism by which protocells were able to take up nutrients from their environment or expel waste. The high stability of phospholipid vesicles also inhibits simple growth mechanisms, namely, incorporation of freshly added phospholipids into the membrane. This inability to grow due to inherent membrane properties would have posed challenges for the evolution of protocell division and replication.

Fatty acid vesicles on the other hand are much less stable and can only form under a relatively narrow range of solution conditions. If the pH is too acidic, the carboxylate head groups become protonated and the fatty acids phase separate. If the solution is too basic, the fatty acids become deprotonated, which favours the formation of micelles instead. Only when the solution pH is attuned to the pK_a of the fatty acids when part of the membrane (which occurs at pH ~8) are fatty acid vesicles capable of forming, wherein the outer and inner hydrophilic surfaces are composed of ~1:1 carboxylate/carboxylic acid head groups that form a stable hydrogen-bonded

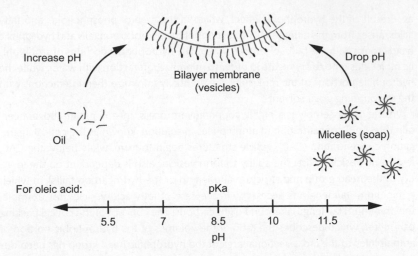

Figure 8.6 The pH dependence of fatty-acid self-assembly into vesicles, micelles, or phase-separation into an oil.

network, see Figure 8.6. Multi-lamellar vesicles composed of multiple membrane layers also commonly form.

The lesser stability of fatty acid vesicles does have advantages, however, for example, making bilayers more permeable to relatively small, charged species while being largely impermeable to large and/or highly charged molecules like RNA and peptides. Growth and division mechanisms for fatty acid vesicles are also more accessible because of their lesser stabilities. They are, however, vulnerable to divalent cations like Ca^{2+} and Mg^{2+}, which form insoluble salts by association with the carboxylate head groups, while phospholipid vesicles are more tolerant of these common cations. This fact will come up again in the context of nonenzymatic template-directed RNA synthesis, which requires divalent metal catalysis, and will be discussed.

Vesicle formation can be assisted by mineral surfaces

The minimal conditions needed for vesicle self-assembly are, in principle, the presence of suitable amphiphiles and water, but geochemically realistic scenarios for this process will almost always involve mineral surfaces of some sort. The surface chemistry of mineral particles has been shown to enhance vesicle formation. For example, silica, apatite, and montmorillonite clay particles have been shown to enhance vesicle self-assembly. Flat mineral surfaces can also be used. In one example, the amphiphile, when exposed to the flat mineral surface, forms a bilayer film, which can then bud-off into individual vesicles as shown in Figure 8.7.

Encapsulation and transport mechanisms

Because large polar molecules like RNA and peptides are essentially incapable of crossing a fatty acid vesicle bilayer, encapsulation cannot occur by passive diffusion from the bulk solution to the vesicle interior. Molecules like RNA, however, can be randomly encapsulated if they are present during initial vesicle formation. Alternatively, wet-dry cycles can also promote encapsulation. During dry-down, vesicles may be disrupted, allowing amphiphiles to form bilayer or multilamellar films, which allow RNA to be sandwiched between the layers. When the mixture is rehydrated, vesicles reform, and some enmeshed RNA becomes encapsulated, see Figure 8.8.

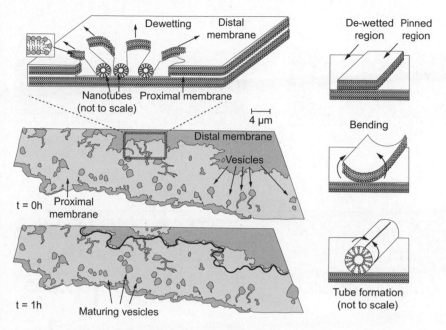

Figure 8.7 Scheme showing surface-assisted vesicle formation. Scheme reproduced from E. S. Köksal et al. (2019). 'Nanotube-Mediated Path to Protocell Formation'. *ACS Nano 13*: 6867–6878.

Note: The lipids first form bilayers on the surface, which then curl into nanotubes and eventually produce budding vesicles.

Transport of small or minimally charged molecules across fatty acid vesicles is thought to occur by a few mechanisms, including the so-called solubility-diffusion model, the transient-pore model, the head-group gated model, and the lipid-flipping carrier model. In all these mechanisms, transport happens as a

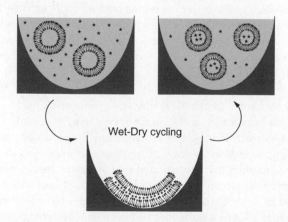

Figure 8.8 RNA encapsulation in fatty-acid vesicles mediated by wet-dry cycles. Scheme reproduced from S. Sarkar et al. (2021). 'Influence of Wet–Dry Cycling on the Self-Assembly and Physicochemical Properties of Model Protocellular Membrane Systems'. *ChemSystemsChem 3*: e2100014.

Note: the stars represent RNA.

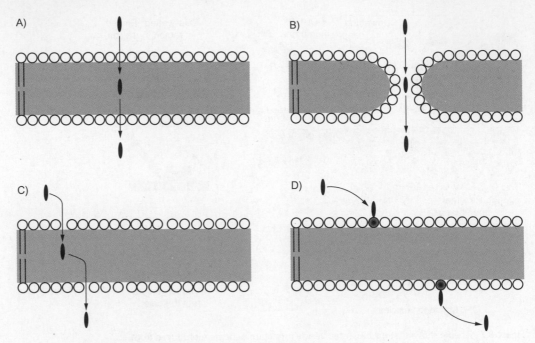

Figure 8.9 Scheme depicting the **A)** solubility-diffusion, **B)** transient pore, **C)** lipid-head-group-gated, and **D)** flip-flop carrier models for transport of substrates across fatty acid vesicle membranes.

consequence of concentration gradients, with molecules diffusing from high to low concentrations, see Figure 8.9.

Transmembrane pH gradients

Electrochemical proton gradients play a key role in extant metabolism, for example, by providing the energy for ATP synthesis via the ubiquitous synthase proteins. Fatty acid vesicles are also capable of forming proton gradients, although they tend to dissipate relatively quickly in comparison to gradients created with phospholipid vesicles. Membrane growth itself, by addition of fatty acids to solutions of preformed vesicles, can produce a pH gradient with respect to the inside and outside of the vesicle. As new fatty acids are incorporated into the outer leaflet of the membrane, half of them must become transferred to the inner leaflet via a flip-flop mechanism. Since flip-flop is much faster for neutral (protonated) fatty acids, these acidic lipids are selectively transferred to the inner leaflet, acidifying the interior, see Figure 8.10.

Proton gradients across vesicle membranes can be generated chemically and photochemically, at least in the case of phospholipids, which serve as proof-of-principle demonstrations that sophisticated proton pumps are not needed to generate transmembrane proton gradients. In one example, it was shown that a small peptide bound to iron could serve as a conduit through which electrons originating from nicotinamide adenine dinucleotide hydrogen (NADH) could be shuttled across the membrane to H_2O_2, an electron acceptor which reacts to form ^-OH, leading to a transmembrane pH gradient. Quinones derived from meteorite extracts as well as polycyclic aromatic hydrocarbons (PAHs) have also been shown capable of shuttling electrons across the membrane, see Figure 8.11.

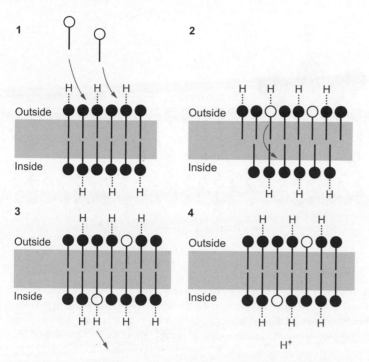

Figure 8.10 Scheme depicting how the addition of external fatty acids to a fatty acid membrane can lead to a pH gradient across the membrane by acidifying the interior. Scheme reproduced from I. Chen, J. W. Szostak. (2004). 'Membrane Growth Can Generate a Transmembrane pH Gradient in Fatty Acid Vesicles'. *PNAS 101*: 7965–7970.

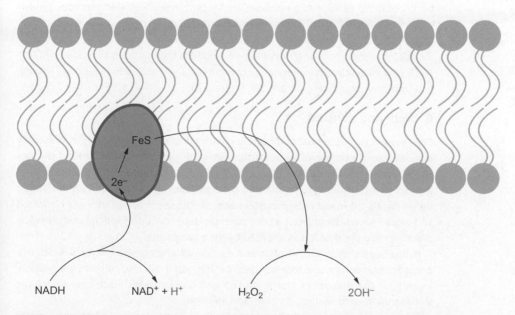

Figure 8.11 Example of how redox chemistry involving a membrane-bound iron-bearing peptide can generate a pH gradient by increasing the alkalinity of the interior. Scheme adapted from C. Bonfio et al. (2018). 'Prebiotic Iron–Sulfur Peptide Catalysts Generate a pH Gradient Across Model Membranes of Late Protocells'. *Nature Cat 1*: 616–623.

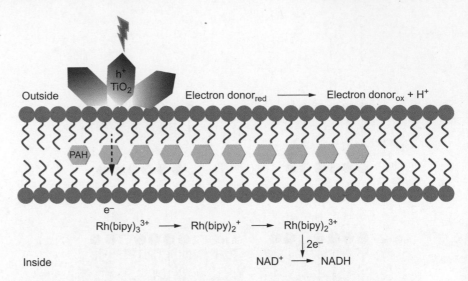

Figure 8.12 Scheme showing how a pH gradient can be generated through the UV photochemistry of TiO_2 mediated by electron transfer through PAHs imbedded in the membrane. Scheme reproduced from P. Dalai, N. Sahai. (2020). 'A Model Protometabolic Pathway across Protocell Membranes Assisted by Photocatalytic Minerals'. *J Phys Chem C 124*: 1469–1477.

Phospholipid vesicles encapsulating ferrocyanide, $[FeCN_6]^{4-}$, can establish a pH gradient photochemically. Upon ultraviolet (UV) illumination, ferrocyanide undergoes ligand exchange, releasing a molecule of ^-CN, substituting its coordination with an H_2O molecule. Cyanide anion is a moderate base, and the interior of the vesicle becomes more alkaline as a result. In another more sophisticated example, photocatalytic minerals, like TiO_2, upon illumination were shown to be capable of transferring electrons into the interior of a phospholipid vesicle using PAHs imbedded in the membrane as electron shuttles. These electrons then went on to reduce NAD^+ to NADH (mediated by a rhenium metal complex catalyst) while simultaneously leading to a transmembrane pH gradient, see Figure 8.12.

Growth and division

Cell division is an essential component of reproduction, a process which is actively managed by a number of proteins that, for example, form a ring around the circumference of the middle portion of the cellular membrane and contracts it until it 'pinches off', resulting in fission that produces two daughter cells. Cell division in the earliest lipid membrane protocells was almost certainly not very sophisticated and would have instead been a phenomenon dependent on spontaneous chemical processes and physical forces afforded by the environment.

It has been shown that growth and division of amphiphilic bilayer vesicles can occur spontaneously, and that vesicle growth by itself can lead to protocell division. In general, the process of vesicle growth and division takes place over four basic steps: growth, deformation, division, and inflation.

There are three basic mechanisms for growth: (*i*) incorporation of amphiphilic/membrane molecules externally from the surrounding aqueous environment, (*ii*) synthesis of constituent membrane molecules internally, or by (*iii*) fusion/competition with other vesicles. Deformation of the spherical shape of the vesicle, i.e. into a more oblong pill-like

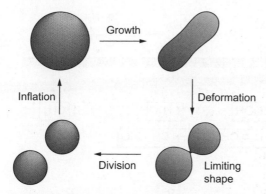

Figure 8.13 Scheme showing vesicle growth and division by addition of amphiphiles (during the growth stage).

shape, is a necessary step prior to division. Deformation can happen, for example, when vesicle growth occurs by external addition of amphiphilic molecules to the membrane. Under the right circumstances, the surface area of the membrane will grow faster than the internal volume, causing the vesicles to 'stretch' rather than expand uniformly. Gentle agitation (e.g. by wind or wave action) can provide the physical forces needed to facilitate division. Membrane tension instabilities can also lead to budding. The final phase is inflation, in which the divided vesicles expand in volume again, see Figure 8.13.

In another noteworthy example of a vesicle growth mechanism, when RNA is encapsulated within fatty acid vesicles, the osmotic pressure results in membrane tension and provides a driving force for vesicle enlargement. Vesicles containing RNA grow by 'stealing' fatty acids from those not containing RNA, alleviating the tension on the membrane. Thus, protocells capable of synthesizing RNA efficiently could have outcompeted those which could not, providing an important selection pressure for Darwinian evolution.

8.4 Prebiotic Fatty Acid and Phospholipid Synthesis

As promising as fatty acid vesicles seem for protocell synthesis, the prebiotic synthesis of fatty acids is still not well understood. The primary challenge to synthesizing fatty acids is the hydrocarbon chains themselves being in a highly reduced state. While formation of a C–C bond between two hydrocarbon fragments is easy to envision using modern synthetic organic reagents and standard S_N2 protocols, it is difficult to envision what the prebiotic equivalent of these reagents could have been. Hence, alternative mechanisms to nucleophilic substitution reactions with sp³ carbons may be necessary.

Fischer–Tropsch and Miller–Urey-type synthesis of fatty acids

The most highly referenced procedures for prebiotic fatty acid synthesis rely on Fischer–Tropsch-type (FTT) mechanisms. FTT reactions typically employ metal catalysts like iron or cobalt, as well as CO and H_2 gases under high pressures and temperatures. The CO and H_2 gases first adsorb to the surface of the metal, which facilitates the reduction of CO to surface-bound CH_3. Another molecule of CO can then insert itself between the CH_3 and the metal surface, followed by another round of reduction and so on, until a long surface-bound hydrocarbon chain is furnished. Release of the hydrocarbon chain from the surface can occur with an accompanying oxidation resulting in the formation of a fatty acid. Relatively long fatty acids can be produced

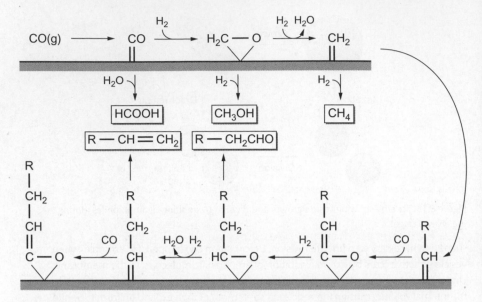

Figure 8.14 Mechanism for FTT fatty acid synthesis occurring on a surface using CO and H_2 as starting materials. Scheme reproduced from R. Partington et al. (2020). 'Quantitative Carbon Distribution Analysis of Hydrocarbons, Alcohols and Carboxylic Acids in a Fischer-Tropsch Product from a Co/TiO$_2$ Catalyst During Gas Phase Pilot Plant Operation'. *J Anal Sci Tech 11*: 1–20.

in this way. It is conceivable that such syntheses could have happened prebiotically in geologic environments rich in H_2 and CO gas in the presence of metallic iron, which, for example, could have been delivered via meteoritic impacts, see Figure 8.14.

Another known mechanism for prebiotic fatty acid synthesis comes from Miller–Urey type experiments. Spark discharges acting on gaseous methane produces hydrocarbons, and in the presence of water can yield carboxylated derivatives. Shorter fatty acids generally result from Miller–Urey-type mechanisms in comparison to FTT, presumably because larger hydrocarbons tend to condense out of the gas phase before they can react further. Branched as opposed to linear hydrocarbon chains are also more prevalent in Miller–Urey-type fatty acid synthesis.

It is worth mentioning that fatty acids can be reasonably abundant in carbonaceous meteorites like the Murchison, and it is thought that a combination of mechanisms leads to their astrochemical production. Chemical extracts of the Murchison meteorite when dissolved in water can even result in the formation of vesicle-like structures.

Iterative aldol condensation and reduction

Aldol condensation followed by reduction is another potential way to generate long-chain hydrophobic hydrocarbons. Repetitive cycles of homoaldol condensation with acetaldehyde can yield long-chain conjugated alkenes, which can be reduced to afford saturated fatty acids. This so-called homoaldolization can take place in alkaline aqueous conditions at room temperature. The fact that the products formed are highly conjugated may provide a driving force for these condensation reactions to occur even in aqueous solution. Reduction of the alkenes to sp^3 hydrocarbons can be mediated via hypophosphite acting as a reducing agent in the presence of nickel metal as a catalyst, conditions of which mimic the corrosion of schreibersite, an iron phosphide mineral found in meteorites, see Figure 8.15.

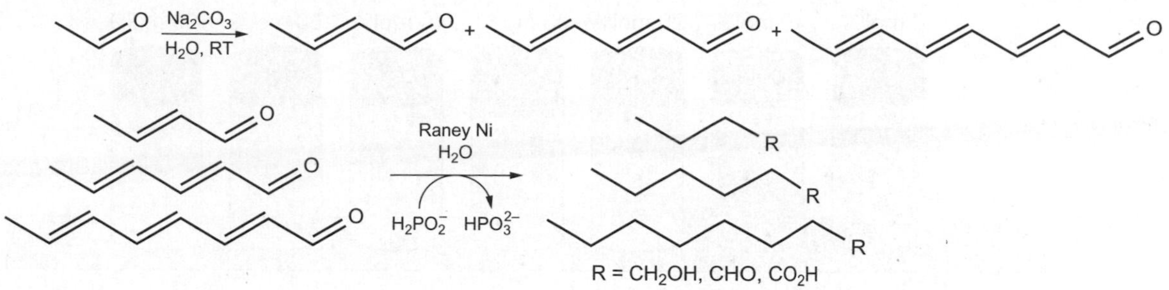

Figure 8.15 Scheme for iterative aldol condensation followed by reduction to fatty acids, alcohols, and aldehydes. For more details, see C. Bonfio et al. (2019). 'Length-Selective Synthesis of Acylglycerol-Phosphates through Energy-Dissipative Cycling'. *J Am Chem Soc 141*: 3934–3939.

Fatty acid esterification and phosphorylation

To synthesize phospholipids resembling biological ones, fatty acids must be esterified with glycerol, followed or preceded by phosphorylation. Simply drying down aqueous mixtures of fatty acids with glycerol can afford glycerol esters. Inclusion of phosphate and a condensation agent (e.g. cyanamide) along with glycerol and fatty acid can yield phospholipids after heating the aqueous phase. It is also possible to preform glycerol phosphate by reaction of glycerol and inorganic phosphate in a (nonaqueous) deep-eutectic solvent or by using an activated form of phosphate, namely, diamidophosphate (DAP, $H_2NPO_2NH_2^-$), whose NH_2 groups become good leaving groups and can be substituted. Note that while glycerol lacks an asymmetric centre, (phospho) esterification at the 1 or 3 positions generates an asymmetric centre at the 2-carbon. As discussed in Chapter 3, bacterial phospholipids employ the D-enantiomers while archaeal phospholipids utilize the L-enantiomers, see Figure 8.16.

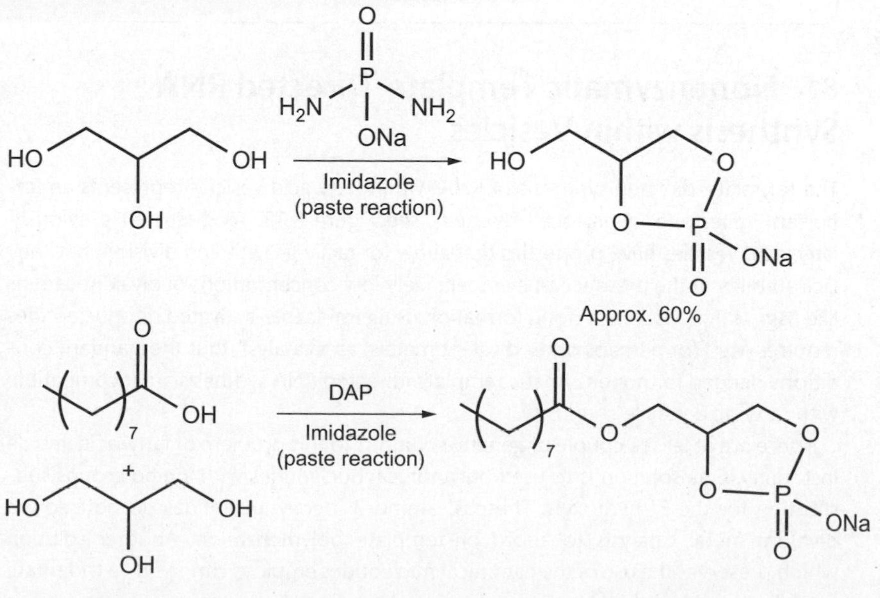

Figure 8.16 Scheme showing diamidophosphate-mediated phosphorylation of glycerol and the formation of a fatty acid glyceride. For more details, see C. Gibard et al. (2018). 'Phosphorylation, Oligomerization and Self-Assembly in Water under Potential Prebiotic Conditions'. *Nature Chem 10*: 212–217.

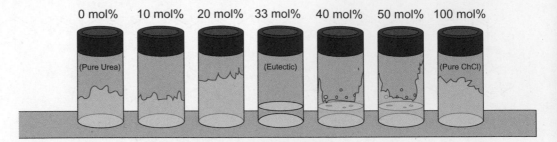

Figure 8.17 Schematic drawing of a deep eutectic solvent (middle vial) formed from choline and urea. Image adapted from B. B. Hansen et al. (2021). 'Deep Eutectic Solvents: A Review of Fundamentals and Applications'. *Chem Rev 121*: 1232–1285.

Deep eutectic solvents

A eutectic system is a homogeneous mixture of two or more compounds that melts at a single temperature that is lower than any of the melting points of the individual components. A deep eutectic solvent is one whose melting temperature becomes so depressed that it forms a liquid at room temperature and pressure. One example of a deep eutectic solvent is the 1:2 mixture of choline chloride (part of the head group of phosphatidylcholine, a biotic phospholipid) and urea. The individual melting points of choline chloride and urea are 302 °C and 133 °C, respectively, but the melting point of the eutectic mixture is 12 °C and is thus a liquid at room temperature that can be exploited as a nonaqueous solvent under standard conditions. Deep eutectic solvents form from mixtures of common, biologically derived mixtures of hydrogen bond donors and acceptors, including simple amino and carboxylic acids. These non-aqueous solvents are beginning to draw attention as potentially prebiotic alternatives to water that can facilitate, among other things, condensation reactions, see Figure 8.17.

8.5 Nonenzymatic Template-Directed RNA Synthesis within Vesicles

The template-directed synthesis of RNA within fatty acid vesicles represents an important milestone in protocell research, see Figure 8.18. As discussed previously, fatty-acid vesicles have properties that allow for facile growth and division, but they lack stability in the presence of even relatively low concentrations of divalent cations like Mg^{2+}. Phosphodiester bond formation using imidazole-activated ribonucleotides requires Mg^{2+} (or perhaps other divalent metals) as a catalyst, thus the standard conditions needed for nonenzymatic template-directed RNA synthesis are incompatible with fatty-acid vesicle chemistry.

There are at least a couple of general solutions to this problem of fatty-acid vesicle instability. One solution is to use noncanonical nucleotides with amino groups substituting for the 3′-hydroxyls. These 3′-amino-3′-deoxynucleotides do not require divalent metal catalysts to afford on-template polymerization. Another solution which preserves the use of the canonical nucleotides employs citric acid—a tridentate ligand—as an Mg^{2+} chelator. Citrate binds to Mg^{2+} in such a way that prevents it from strongly interacting with fatty-acid head groups, hence preserving vesicle stability,

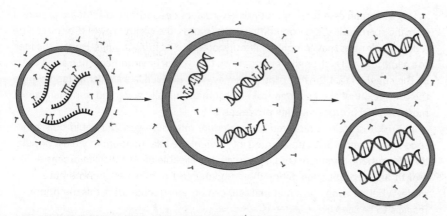

Figure 8.18 General scheme for nonenzymatic template-directed synthesis of RNA within fatty acid vesicles. For more details, see S. S. Mansy et al. (2008). 'Template-Directed Synthesis of a Genetic Polymer in a Model Protocell'. *Nature 454*: 122–125.

while still affording open coordination sites for catalytic activity. Simple addition of citrate to the reaction mixture allows template-directed synthesis of RNA within fatty-acid vesicles to be achieved. Moreover, the activated (as well as spent) ribonucleotide monomers are capable of passing through the vesicle membrane, while the template and polymerized products remain compartmentalized inside. Citrate also suppresses the Mg^{2+}-catalysed hydrolysis of RNA.

8.6 An Introduction to Synthetic Biology

Our discussion of protocells thus far has implied the use of what is known as a bottom-up approach, based on the compounds that could plausibly have been supplied by prebiotic chemistry, but there is also a top-down approach to constructing protocells in the lab. As you might have already inferred, the bottom-up approach is based on constructing complex chemical systems based on what one hypothesizes the environment may have made abundant, while the top-down approach is based on deconstructing modern cells and stripping their complexity until they possess only the minimum information necessary to be capable of carrying out fundamental life processes, i.e. metabolism, reproduction, and evolution.

This manner of constructing artificial protocells is part of a larger field known as synthetic biology. According to the National Human Genome Research Institute, synthetic biology is 'a field of science that involves redesigning organisms for useful purposes by engineering them to have new abilities', or in the case of origins-of-life research, the minimum set of abilities needed to be considered 'alive', resulting in a synthetic organism known as a minimum cell.

The top-down approach to constructing a minimum cell has seen a number of successes, especially in recent years and has provided insight into how protocells on early Earth may have regulated essential life processes. As briefly discussed in Chapter 3, much of this work has centred around the parasitic bacterium, *Mycoplasma genitalium*, which with only 517 genes is one of the simplest organisms known. Researchers were able to deduce through gene knock-out experiments that only about half of these genes are necessary. In fact, the number of genes can be reduced even further, as long as certain essential compounds like nucleotides and amino acids are provided from the surrounding environment. In other words, the complexity of the cell can be reduced as long as that complexity is 'outsourced' to its surroundings. Hence, the idea of a minimum cell is strongly dependent on the type of environment such a cell exists in. This suggests that the first protocells may have evolved in a similar fashion, where the complexity of the geochemical environment initially governed and maintained most essential protocellular life processes.

In 2016, researchers reported that an entire minimum genome could be synthesized 'from scratch' and transplanted into a viable cellular medium to yield a population of synthetic minimum cells. These cells are capable of essential life processes like growth and reproduction. Since then, updates to the minimal genome have been made. Whether even more minimal cells can be constructed using this methodology remains to be seen.

For the first time in human history, creating artificial life in the lab is in the realm of possibility. While the bottom-up approach to constructing a population of protocells capable of self-sustained growth, reproduction, and evolution still has some way to go, many scientists now think that this is an achievable goal in the not-too-distant future. The 21st century may indeed see this fundamental goal of prebiotic chemistry and science in general realized. What protocell in the future might you design?

8.7 Summary

- A protocell is any theoretical or experimental model involving a cell-like compartment.

- Compartments are necessary for life as we know it, partly because they allow for Darwinian evolution of complex adaptive traits.

- Vesicles are supramolecular assemblies containing an aqueous interior separated from the bulk solution by one or more bilayers, which under the right conditions spontaneously self-assemble.

- Double-chain phospholipids form more stable vesicles than single-chain fatty acids. This greater stability could counterintuitively pose problems for hypothetical early protocells.

- The most well-studied prebiotic fatty acid syntheses rely on Fischer–Tropsch-type (FTT) mechanisms. FTT typically employs metal catalysts, and CO and H_2 gases under high pressures and temperatures.

- Synthetic biology involves redesigning organisms for useful purposes by engineering them to have new abilities, or in the case of origins-of-life research, the minimum set of abilities needed to be considered 'alive'.

8.8 Exercises

1. What are contemporary cell membranes composed of?
2. What are some major potential prebiotic sources of amphiphiles?
3. What is a protocell?
4. What is a minimal cell?
5. What do you think are the most fundamental advantages of cellularity?

8.9 Suggested Reading

1. David C. Krakauer, David Deamer, Liaohai Chen, Mark A. Bedau, and Steen Rasmussen (eds). (2022). *Protocells: Bridging Nonliving and Living Matter*. United States: MIT Press.

2. Daniel G. Gibson, et al. (2010). 'Creation of a Bacterial Cell Controlled by a Chemically Synthesized Genome'. *Science 329* (5987): 52–56.

3. Can Xu, Shuo Hu, and Xiaoyuan Chen. (2016). 'Artificial Cells: From Basic Science to Applications'. *Mat Today 19* (9): 516–532.

4. Jacob A. Vance and Neal K. Devaraj. (2021). 'Membrane Mimetic Chemistry in Artificial Cells'. *J Am Chem Soc 143*: 8223–8231.

5. Nicolas Martin and Jean-Paul Douliez. (2021). 'Fatty Acid Vesicles and Coacervates as Model Prebiotic Protocells'. *ChemSystemsChem 3*: e2100024.

Glossary

Abiotic any physical or chemical process occurring in natural environments driven by mechanisms that do not involve biology, e.g. microorganisms.

Accretion the process of planet formation, either rocky or gaseous, by the accumulation of dust and gas from the disc located around a newly formed star.

Aerosol small liquid or solid particles suspended in a gas, e.g. the atmosphere.

Aldol addition a reaction that combines two carbonyl compounds (aldehydes or ketones) to form a new β-hydroxy carbonyl compound. This is one of the main reactions taking place in the formose reaction.

Aldoses a sugar which is an aldehyde. Can also refer to their corresponding intramolecular (cyclic) hemiacetals.

Amino acids organic molecules that contain at least one primary amine group (NH_2) and one carboxyl group (COOH). This amine need not be alpha to the carboxyl group, although this is the case for proteinogenic amino acids.

Anomers diastereoisomers of glycosides, hemiacetals or related cyclic forms of sugars, or related molecules, which differ in stereochemical configuration only at C-1 of an aldose or the carbonyl carbon of a ketose.

Apatite a group of phosphate minerals, which are typically insoluble in water. Although a potential source of phosphate on early Earth, their insolubility inhibits their ability to form organic phosphates.

Archean the second major period in geological history preceded by the Hadean and followed by the Proterozoic. The geologic record reveals life was well-established by the time of the Archean.

Asymmetric centre an atom having a spatial arrangement of bonded atoms or lone pairs which is not superposable with its mirror image. In the case of carbon, an asymmetric centre can only be a tetrahedral carbon bonded to four different functional groups or atoms.

Autocatalysis catalysis by one or more of the products of a reaction. Autocatalysis occurs when the starting materials of a reaction produce two or more copies of at least one of those starting materials.

Autotroph an organism capable of biosynthesizing all of its necessary material from carbon dioxide as the only carbon source.

Cannizzaro reaction a chemical reaction which involves the disproportionation of two molecules of an aldehyde to give a primary alcohol and a carboxylic acid.

Carbonyl migration a type of tautomerization reaction involving the transposition of a carbonyl group (aldehyde or ketone) to another site in the same molecule. In the case of sugars, carbonyl migration can occur through enediol(ate) intermediates or hydride transfer.

Cell membrane the boundary that encases a cell, providing a semipermeable barrier that separates the cell from its surroundings.

Cell wall a rigid external layer that surrounds some types of cells and has structural, protective, and functional roles. Cell walls are of such structural complexity that they are typically ignored in origins-of-life studies.

Central dogma the notion that biological information flows only in one direction, from DNA, to RNA, to proteins. It is now known that information can also flow from RNA to DNA via reverse transcriptases.

Chemical evolution the set of naturally occurring (geochemical) reactions likely having taken place over geologic timescales that led to the first living systems from abiotically synthesized molecules on the early Earth.

Chemoton (short for 'chemical automaton') refers to an abstract model for the fundamental unit of life introduced by Hungarian theoretical biologist Tibor Gánti.

Chimeric template a nucleic acid template which contains more than one type of monomer (e.g. both RNA and DNA nucleotides in the same oligomer or polymer).

Chondrite primitive undifferentiated stony meteorites consisting of fine-grained dust particles, surrounding roughly millimetre-sized, round inclusions (chondrules) and other types of metal-rich particles and mineral grains.

Coacervate a condensed liquid-like phase, usually composed of oppositely charged polymeric molecules (e.g. nucleotides and positively charged peptides) formed by liquid-liquid phase separation typically from water.

Condensation reaction a reaction (usually involving more than one step) that produces a single main product from two or more reactants (or reactive sites within the same molecule) and is typically accompanied by the release of a water molecule.

Coronal mass ejection (CME) an ejection of a large mass of plasma from a star's corona powered by magnetic fields. CMEs are typically associated with solar flares and other forms of solar activity and can result in significant atmospheric chemistry.

Crust the outermost solid shell of a rocky planet (or satellite like the Moon). The crust may be composed of rock and/or ice.

Cyanohydrins alcohols substituted by a cyano (nitrile) group wherein typically the cyano and hydroxy groups are attached to the same carbon atom. Cyanohydrins in prebiotic chemistry are derived from the addition of hydrogen cyanide to aldehydes or ketones.

Darwinian evolution the theory that species originate by descent, with variation, from parents (or single cells), through the natural selection of those individuals best adapted for reproduction; also known as Darwinism. The origin of Darwinian evolution is a key question in origins-of-life studies.

Deep time vast, extremely remote periods of natural history not easily compared to human timescales.

Depsipeptide a peptide in which one or more of its amide groups are replaced by an ester group. Depsipeptides have both peptide and ester linkages.

Diastereomers stereoisomers (i.e. compounds with the same molecular formulae and connectivity between atoms) that are not related as mirror images.

Differentiation separation of the different constituents of planetary materials resulting in the formation of distinct compositional layers. Denser material tends to sink into the centre and less dense material rises toward the surface.

Electric discharge the release and transmission of electricity by the application of an electric field typically through a gaseous medium, e.g. the atmosphere. Lightning is a common form of electric discharge.

Enantiomer one of a pair of molecules, which are mirror images of each other and are non-superposable. The fact that they are mirror images implies that they have the same molecular formula and same connectivity between their constituent atoms.

Eukaryote one of the three domains of life distinguished from bacteria and archaea at the morphological and molecular levels. All members of the eukarya have a nucleus and are further distinguished from bacteria and archaea by a more complex cellular organization.

Eutectic phase a homogeneous mixture that has a melting point lower than the individual components that make up the mixture.

Extraterrestrial of or from outside the solid Earth and its atmosphere and oceans.

Extremophiles organisms that thrive in ecosystems characterized by one or more physical parameters (e.g. pH, temperature, salinity) so extreme that they are close to the known limits of life.

Fatty acids aliphatic monocarboxylic acids typically derived from animal or vegetable fat, oil, or wax. The term can also be used to refer to all acyclic aliphatic carboxylic acids regardless of whether they are obtained from biological sources.

Fidelity of copying the accuracy of template-dependent nucleic acid copying, in particular DNA or RNA replication. The fidelity of nonenzymatic replication tends to be substantially lower than its enzymatic counterpart.

Furanoses cyclic hemiacetal forms of monosaccharides (aldoses or ketoses) in which the ring is five-membered having a tetrahydrofuran skeleton.

Geologic hotspot columns of hot buoyant mantle that originate from a thermal boundary layer deep in the Earth; also known as a mantle plume.

Glycosidic bond a type of bond that joins a carbohydrate (sugar) molecule to another group, e.g. another sugar or a nucleobase.

Hadean the earliest eon of geological time, extending from accretion of the Earth (4.567 Ga) to the formation of the earliest known rocks in the Archean at ~4.0 Ga.

Haze a suspension of fine liquid droplets or solid particles in an atmosphere, which blocks or refracts incident light (see also aerosol).

Heterotrophic hypothesis a hypothesis for the origin of life, first proposed by Oparin and Haldane, that suggests early organisms depended on abiotically synthesized organic molecules for their structural components and as an energy source.

Heterotrophic organisms any organism that needs organic compounds as a carbon source for the synthesis of its own cellular components. According to their source of energy, organisms can be classified as photoheterotrophs or chemoorganotrophs.

Homeostasis the maintenance of the conditions in which a system is viable, despite external perturbations or differences with its environment.

Homologation any chemical reaction that converts a set of reactant(s) into the next member of the homologous series, i.e. a group of compounds that differ by a constant chemical unit, e.g. CH_2O in the case of sugars.

Hydrated electron a free electron solvated in water with a short lifetime that can serve as a reducing agent.

Hydration addition of water or of the elements of water (i.e. H and OH) to a molecular entity.

Hydrolysis the reaction of a molecule with a water molecule, or with a hydroxide ion, involving the rupture of one or more bonds that can result in splitting of the molecule into two smaller compounds.

Isomers molecules that have the same atomic composition (molecular formula) but different connectivities between their constituent atoms or different stereochemical configurations, and hence they have different physical and/or chemical properties.

Kaolinite a clay mineral, with the chemical composition $Al_2Si_2O_5(OH)_4$. It is a layered silicate mineral composed of silica (SiO_4) and alumina (AlO_6) sheets.

Ketose a sugar that exists as a ketone including their intramolecular (cyclic) hemiacetals.

Kiliani–Fischer synthesis a method for synthesizing monosaccharides, that relies on addition of hydrogen cyanide to form cyanohydrins followed by reduction. In prebiotic chemistry, reductions can be carried out by hydrated electrons, for example.

Kuiper belt the region in the Solar System beyond the orbit of Neptune that contains icy bodies orbiting the Sun.

Late veneer the late accretion of asteroidal or cometary material to the terrestrial planets. Differentiation results in iron and nickel segregation during core formation and left the mantle of the Earth depleted in siderophile (iron-loving) elements, notably platinum-group elements. The modern abundances of these elements in Earth's mantle greatly exceed the level expected after differentiation, and hence, were delivered afterwards at a later stage.

Late-heavy bombardment a hypothesized elevated frequency of asteroidal and other small body collisions that affected the inner Solar System, e.g. the Earth, between 4.0 and 3.8 billion years ago.

Leaving group an atom or functional group that becomes detached from a reactant during a reaction allowing for the formation of a new bond.

Life (the National Aeronautics and Space Administration (NASA)'s working definition): a self-sustaining chemical system capable of Darwinian evolution.

Lipid a loosely defined term for substances typically of biological origin that are soluble in nonpolar solvents and often are components of cell membranes (e.g. fatty acids and phospholipids).

LUCA (last universal common ancestor) the most recent ancestor from which all currently living species evolved.

Magma ocean the surface condition during the formation of primitive rocky planets wherein molten rock composes the uppermost planetary layer, and which has a substantial influence over the composition of the primitive atmosphere through controlling outgassing.

Mantle an intermediate layer in a differentiated planet or small Solar System body typically composed of silicates and metal oxides, which separates the body's core and crust.

Metabolism the sum of chemical processes living organisms use to convert raw feedstocks into energy and cellular components.

Meteorite a remnant of an asteroid or comet which manages to enter the atmosphere of a planet and survive impact with the surface.

Micropalaeontology the study of the preserved remains of microscopic organisms.

Montmorillonite a common smectite phyllosilicate mineral, or clay, with the formula $(Na,Ca)_{0.3}(Al,Mg)_2(Si_4O_{10})(OH)_2 \cdot nH_2O$. It has been used as a standard common clay mineral in prebiotic chemistry studies of the interactions of organics with mineral surfaces in aqueous environments.

mRNA (messenger RNA) RNA molecules generated from the process of transcription from DNA templates and typically read out into peptides.

Neighbouring group participation the interaction of a reaction centre with a lone pair of electrons in an atom or the electrons present in a sigma bond or pi bond present in the same molecule but is not conjugated with the reaction centre. A rate increase due to neighbouring group participation is known as 'anchimeric assistance'.

Nucleoside the collective structure formed from a nucleic acid base and a sugar.

Nucleosynthesis the synthesis of new atom types via high-energy processes. Important loci of nucleosynthesis include the Big Bang, stellar fusion, and supernova events.

Nucleotide a molecule composed of a nucleic acid base, a carbohydrate, and a phosphate moiety.

Oort cloud a cloud of icy bodies believed to surround the Sun at a distance of 2000–200,000 AU existing as spherical and disc-shaped regions. The Oort cloud represents material leftover from Solar System formation that defines the edge of the Solar System.

Oxidation state also commonly referred to as oxidation number, is the hypothetical charge of an atom if all of its bonds to different atoms were fully ionic. This is rarely the case, but a useful formalism for chemistry.

Peptide an oligomer or polymer derived from two or more amino acids linked by peptide bonds.

Peptide bond an amide bond formed from the dehydrative linkage of two amino acids.

Phosphodiester linkages a bonding configuration formed from the dehydrative linkages of a phosphate group and two alcohol groups, creating two phosphate esters.

Phospholipid a lipid containing a covalently bound phosphate moiety typically formed from fatty acids, glycerol, and phosphate.

Phosphorimidazolides model compounds in which an imidazole moiety is covalently bound to a phosphate group through N–P linkages, which generally enable kinetically rapid study of organic polymerization processes, e.g. nonenzymatic template-directed RNA synthesis.

Plate tectonics the process by which large conglomerates of planetary crust move by virtue of convective internal planetary heat production and cooperative materials properties.

Prebiotic adjective applied to processes which precede the advent of biology, and contribute to the emergence of biology. To be contrasted with 'abiotic' which describes processes that occur in the absence of biology but do not necessarily assist in its development.

Primary atmosphere the initial atmosphere a planet receives or develops as a result of accretion or magma ocean processes.

Primary metabolite a molecule commonly involved in the normal growth of a large variety of organisms, to be contrasted with secondary metabolites only made by specific organisms which are not necessary for normal growth, development, or reproduction.

Primary structure the monomer sequence of a polypeptide or polynucleotide.

Progenote the hypothetical primitive organism or group of organisms which contained all of the common attributes observed in modern biology.

Prokaryote single-celled organisms of the Eubacteria and Archaebacteria which do not contain a cell nucleus.

Protocell a hypothetical or model cell-like structure which is capable of carrying out some of the functions or processes of a modern cell.

Protoplanetary disc a rotating circumstellar disc of dense gas and dust surrounding a young, newly formed star.

Purine an aromatic nitrogen heterocycle ring system consisting of a fused imidazole and pyrimidine ring. The nitrogenous bases adenine and guanine of DNA and RNA are purines.

Pyranose cyclic hemiacetal forms of monosaccharides in which the ring is six-membered, i.e. a tetrahydropyran skeleton.

Pyrimidine a six-membered aromatic nitrogen heterocyclic ring system with N atoms in the 1 and 3 positions. The DNA and RNA bases cytosine, thymine, and uracil are pyrimidines.

Quaternary structure the arrangement of a protein which itself is composed of smaller protein subunits arranged in a complex.

Radical an atom, molecule, or ion that has at least one unpaired valence electron. These unpaired electrons make radicals highly chemically reactive.

Retro-aldol cleavage reactions the inverse reaction of an aldol addition resulting in two carbonyl groups (aldehyde or ketone).

Ribosome a large multi-subunit protein-nucleic acid complex, which carries out coded protein synthesis in all known terrestrial living organisms. All ribosomes consist of a large and small subunit, and the RNA portion is responsible for catalysing peptide bond formation.

Ribozyme a catalytic RNA oligomer. The RNA in RNAse P was the first naturally occurring (biogenic) ribozyme identified. Many more are now known (including the ribosome), and synthetic ribozymes are now routinely made in the laboratory.

RNA a biopolymer consisting of repeating ribonucleotides joined via phosphodiester bonds.

RNA world hypothesis the idea that before the evolution of the currently pervasive DNA-RNA-protein biochemistry, organisms depended mainly on RNA for both information storage and catalysis via ribozymes. According to this model, both DNA and protein are outgrowths of primitive RNA biochemistry.

rRNA (ribosomal RNA) the RNA component of the ribosome, which accounts for ~65 per cent of ribosomal mass. Prokaryotic ribosomes typically contain three rRNA molecules, while eukaryotic ribosomes contain four.

Second genetic code the set of structural information, which allows tRNA molecules to efficiently bind mRNA in ribosomal active sites. It may also relate to the degree of discrimination amino acyl tRNA synthetases express.

Secondary atmosphere the atmosphere, which planets host after their accretional atmospheres are lost. Secondary atmospheres typically derive from outgassing of volatiles from planetary interiors.

Secondary metabolite a generic term for compounds an organism produces, which are not directly involved in their own growth, development, and reproduction. Secondary metabolites are generally involved in ecological interactions.

Secondary structure the way RNA or protein polymers self-aggregate to form defined structures including helices and sheets.

Self-assembly a process in which a disordered system of pre-existing components forms an organized structure or pattern as a consequence of specific, local interactions among the components themselves, without external direction.

Serpentinization the hydration and metamorphic transformation of minerals such as olivine and pyroxene to produce serpentinite, a mineral group including lizardite, chrysotile, and antigorite, all having approximately the formula: $Mg_3(Si_2O_5)(OH)_4$ or $(Mg^{2+}, Fe^{2+})_3Si_2O_5(OH)_4$. Hydrogen is an important by-product of serpentinization, and the reducing conditions created by serpentinization are implicated in both abiotic organic synthesis in hydrothermal environments and the origins of life.

Shock synthesis term applied to the formation of chemical bonds during extreme and transient conditions of pressure and temperature such as may occur during meteor impacts.

Siderophile An element or compound, typically transition metals, which tend to dissolve in molten iron phases, and thus becomes sequestered in planetary cores during planetary differentiation.

Solar wind the stream of charged particles emitted by the Sun's corona. This stream is composed of electrons, protons, and alpha particles with kinetic energies between 0.5 and 10 keV. With respect to prebiotic chemistry, the solar wind is important as it may drive atmospheric chemical reactions, which produce reactive species important for chemical evolution.

Strand-inhibition problem refers to the problem that while template-directed synthesis depends on the reversible association of activated monomers on a template, as the synthesized template-complementary oligomer elongates, it no longer dissociates from the template sufficiently to allow for new rounds of template-directed synthesis, and thus template-directed synthesis typically fails to display autocatalytic turnover.

Strecker reaction the hydrolytic reaction of an aldehyde or ketone with cyanide and an amine to produce an α-amino acid, proceeding through α-aminonitrile and α-aminoamide intermediates.

Stromatolites layered sedimentary formations created by shallow water photosynthetically based microbial communities. These mats build up layer-by-layer over time. Although rare today, fossilized stromatolites provide important records of ancient terrestrial life.

Submarine hydrothermal vents locations where seawater becomes entrained via convection with emerging magma at spreading centres along mid-ocean ridges. The interaction of seawater with mantle mineralogy may produce significant disequilibrium chemistry, which has been hypothesized to have been significant for abiogenesis and the emergence of primitive life.

Sugar a generic name for sweet tasting monosaccharides or small oligosaccharides. Glucose and ribose are important common biochemical sugars.

Systems chemistry the study of the networks of interacting molecules, which may create new functions from sets of molecules that lead to different hierarchical levels and emergent properties. Systems chemistry is thought to be important for the origin of life as it describes how sets of simple molecules collectively manage to perform novel selectable functions.

Tautomerization refers to the process of interconversion of tautomers.

Template a molecular species, which presents information that can be copied, e.g. single-stranded RNA or DNA.

Template-directed synthesis a synthetic process involving reversible covalent or noncovalent bonding interactions between activated monomers which are organized by a template into reactive geometries to create specific covalent bond formation. Nucleic acid copying generally depends on template-directed synthesis, and this mechanism has been explored in prebiotic chemistry to understand the origin of nucleic acid replication.

Tertiary structure the three-dimensional shape of a protein, determined by how the primary structure guides the formation of secondary structural motifs, and structural domains. The interactions of sidechains within proteins help determine its tertiary structure.

Transamination the transfer of an amino group from one molecule to another. In biochemistry this is especially common between an amino acid to a ketoacid.

Transcription the synthesis of mRNA transcripts from DNA sequences, which act as a template.

Transesterification the chemical exchange of an esterified group with other alcohol functional groups forming a new ester.

Transesterification reactions may be catalysed by acid or base catalysts and are especially important in the context of RNA splicing.

Translation the conversion of nucleic acid information into peptide information, as occurs using the modern biological translation apparatus, in which mRNA messages are read out into coded peptides.

tRNA (transfer RNA) small, highly modified RNA polymers typically ~76 nucleotides in length which self-assemble into cloverleaf secondary structures, then fold into L-shaped tertiary structures. In their charged forms, they carry an amino acid and adapt the genetic code into a form which allows for readout by the ribosome.

Universal tree of life a generic name for phylogenetic reconstructions of all life on Earth typically based on similarities among rRNA genes. Such trees typically have no real 'root', but are able to effectively relate more recently evolved organisms.

Vesicle a lipid bilayer typically including a liquid (aqueous) interior that differentiates it from a micelle.

Watson–Crick base-pairs purine-pyrimidine base-pairing motifs which use canonical DNA base-pairing, e.g. two hydrogen bonds between A and T, and three between C and G.

Zircons zirconium silicate minerals which form under high temperature and pressure conditions, and which are unusually resistant to chemical weathering, and thus preserve information about the conditions of their formation. Microscopic zircon inclusions have provided information about when life may have first appeared on Earth.

Index